Scaling Issues and Design of MEMS

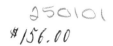

250101

$156.00

Scaling Issues and Design of MEMS

Salvatore Baglio
University of Catania, Italy

Salvatore Castorina
SYNAPTO, Catania, Italy

Nicolò Savalli
University of Catania, Italy

John Wiley & Sons, Ltd

Other Wiley Editorial Offices

John Wiley & Sons Inc., 111 River Street, Hoboken, NJ 07030, USA

Jossey-Bass, 989 Market Street, San Francisco, CA 94103-1741, USA

Wiley-VCH Verlag GmbH, Boschstr. 12, D-69469 Weinheim, Germany

John Wiley & Sons Australia Ltd, 42 McDougall Street, Milton, Queensland 4064, Australia

John Wiley & Sons (Asia) Pte Ltd, 2 Clementi Loop #02-01, Jin Xing Distripark,
Singapore 129809

John Wiley & Sons Canada Ltd, 22 Worcester Road, Etobicoke, Ontario, Canada M9W 1L1

Wiley also publishes its books in a variety of electronic formats. Some content that appears
in print may not be available in electronic books.

Anniversary Logo Design: Richard J. Pacifico

British Library Cataloguing in Publication Data

A catalogue record for this book is available from the British Library

ISBN 978-0-470-01699-2 (HB)

Typeset in 10.5/13pt Sabon by Integra Software Services Pvt. Ltd, Pondicherry, India
Printed and bound in Great Britain by TJ International, Padstow, Cornwall
This book is printed on acid-free paper responsibly manufactured from sustainable forestry
in which at least two trees are planted for each one used for paper production.

Contents

Preface

The concept of scaling is fundamental in engineering design.

This book deals with the main issues in designing microelectro-mechanical systems (MEMS), taking into account the scaling item. This text includes a wide view of various types of MEMS. Moreover, the very important theme of energy supplies in MEMS has been accurately faced. Details of the design of complete MEMS are also reported, referring to the case of colonies of microrobots.

The book explores both theoretical aspects of MEMS models and experimental prototype validation. Analytical, numerical and experimental tools are widely discussed in order to give to the reader a completely clear view in understanding the main topics: to design by using 'scaling' techniques.

The topics discussed are related to the long experience of the authors in designing and in the realization of MEMS devices; the reported study is, in my opinion, at the top level in the considered area of research, moreover it establishes powerful tools to conceive emerging MEMS.

The efforts of the authors have led them to achieve a work where clearness and scientific exactness reached a perfect synchronism!

An accompanying website can be found at http://wiley.com/go/baglio_scaling

Luigi Fortuna
Catania
Italy

Introduction

Accurate modelling and design of microelectromechanical systems (MEMS) cannot be adequately pursued without considering the large number of system interactions between micromechanical parts, many physical and chemical principles, analogue and digital circuits. Many research groups worldwide are effectively making progress in the computing-aided design (CAD) area, to improve significantly quality and design time. Moreover, the academic and industrial target is the definitive integration of micromechanical models with system level simulations.

Despite the fact that such a vast 'microworld' has also gained much space in several basic and advanced, academic courses worldwide, it is necessary to adopt a gradual approach in explaining such a significant quantity of interconnected concepts. Scaling issues of micromechanical systems hold an important role in this context, since almost all classical physical laws must be reviewed when the dimensions scale towards the micro- or nanoscale.

For this reason, while it is important to give a general view of MEMS designing and realization processes, each specific case, referring to the device typology, the used technology, as well as operating conditions (sensing or actuating purposes), and conditioning or driving circuits, must be accurately examined.

Scaling effects on MEMS can be in fact discussed inherently to modelling issues, designing issues, fabrication and micromachining issues. Then, even providing a complete view of scaling issues for these systems would take several efforts and involve many different competences over a relatively long time. One should start from modelling aspects, i.e. mechanical modelling, then continue with etching concepts and issues, i.e. those related to wet-etching procedures, and finally highly complex modelling or systems level simulation of such systems, including circuits.

Many research groups worldwide are not so far away from successfully realizing such aimed software supports, which could significantly improve modelling and designing efficiency in this field.

The aim of this book is that of providing a specific, somehow complementary, theoretical support to the thematic of scaling issues for microelectromechanical systems (MEMS). To do this, several different aspects inherent to the miniaturization of sensors and actuators are proposed in the various chapters, focusing on a general target that is the scaling of an autonomous microsystem.

For this reason, the content of this book covers scaling issues of devices that could be equipped on a microrobot. In this field, it is mandatory to consider scaling issues of energy sources, scaling of microactuators (which can permit the system moving or performing active tasks) and scaling of microsensors (for obtaining information on the system's status itself, or for gathering information on the surrounding environment).

As one can easily realize, also in this case it is not possible to examine all of the possible configurations for autonomous microsystems, due to the large spectrum of potential solutions, neither to define a general paradigm for modelling or designing such kinds of microelectromechanical devices. In fact, several different choices can be made in individuating optimal solutions for energy sources (thermal, electrostatic/capacitive, photo-thermo-electric, battery-powered, etc.), types of actuators (thermal actuators, electrostatic/capacitive, magnetic, piezoelectric, etc.) and particularly sensors.

For this reason, the choice of these authors was that of reporting general considerations of importance for scaling issues of MEMS and then discussing some alternative solutions for all of the three main aspects that always have to be taken into account.

Starting from Chapter 1, where a general discussion on the scaling issues of MEMS is given and related to real-device realizations, the following chapters report specific theoretical and practical issues on the scaling of actuators, sensors, energy sources and finally autonomous MEMS, e.g. microrobots.

In particular, in Chapter 2 the scaling of thermal actuators is discussed inherently to multilayer cantilever beams realized in a CMOS standard technology. In Chapters 3 and 4, the scaling of temperature and relative humidity sensors and magnetic (inductive) sensors are discussed, respectively. The vast class of micromechanical sensors is instead analysed in Chapter 5, with reference to the comparison of performance obtained with the same device realized by means of two different technologies. Importantly, in Chapter 6, the scaling issues of energy sources are considered with reference to innovative photo-thermo-mechanical and photo-thermo-electric energy supply strategies, and the combined use of both strategies is taken into account to investigate the efficiency.

Chapter 7 deals with the technologies and architectures for autonomous MEMS microrobots and reports the description of a real autonomous microsystem with a photo-thermal actuation strategy. In Chapter 8, some considerations on the non-easy ulterior moving towards the nanoscale are reported.

Finally, examples of scaling effects analyses are reported in Chapter 8, for microcantilever beams composed of two layers, together with the technical description of an open-source code realized through Matlab 6.5 for the common typologies of micromechanical devices. In particular, the aim of creating such open software instruments has been that of providing a rapid way to appreciate the effects of scaling on common microdevices that can be also operated as actuators (interdigitated combs) and sensors (mechanical plates, which could be operated as accelerometers and interdigitated combs). A user-friendly interface allows us to choose the adopted technology, the geometrical features of mechanical parts (as sustaining springs) and to evaluate the effects of scaling on the performance of devices. The implemented analytical models include static and pseudo-dynamic analysis of MEMS that are classically equivalent to second-order mass-spring-damper mechanical systems.

Since such software support is conceived to be developed and extended through the help of academic students that are introduced to MEMS, some suggestions on guidelines to be followed are provided to create interesting cultural, and therefore technical exchanges with the collaboration and supervision of authors.

1

Scaling of MEMS

1.1 INTRODUCTION TO SCALING ISSUES

The evolution of microelectronic devices has been characterized by the scaling of their characteristic feature size towards smaller dimensions. The reasons for such a scaling trend are the continuous research of better processing capabilities, which means a higher number of smaller transistors on the chip. Integrated circuits with typical feature sizes in the submicrometre range are currently fabricated and commercialized, and research is pushing such feature sizes towards the regime of a few tenths of nanometres, and even smaller.

Such an amazing revolution has not been limited only to purely electronic devices. In fact, the huge potential for lowcost and large-scale fabrication of semiconductor–microelectronics technologies has represented a very powerful and flexible platform for the conception and realization of miniaturized micromechanical structures, thus opening the way to the realization of miniaturized sensors and actuators. To these authors' knowledge, one of the earlier microelectromechanical systems date back to the end of the 1980s, at the 'Transducers' meeting in Tokyo, 1987, when the first examples of micromachined structures were presented (Gabriel, 1998).

Small mechanical features can be produced with many techniques, and several applications are well consolidated (just think of mechanical wristwatches). In such a sense, MEMS are not a novelty. What is really innovative, instead, is the fabrication process, which makes use of the

Scaling Issues and Design of MEMS S. Baglio, S. Castorina and N. Savalli
© 2007 John Wiley & Sons, Ltd

same technologies and facilities used to produce integrated circuits. This will translate into two of the key characteristics of MEMS: lowcost, large volume batch fabrication and integration of mechanics and electronics on the same semiconductor substrate.

Having both mechanical structures and electronics, realized together on the same chip, means that simple mechanical features can be suitably functionalized and controlled in their operation by means of electronics, and that electronics can process information coming from mechanical parts. In other words, if one thinks of mechanical features on a chip as sensors and actuators, then the space of the signals the electronics can deal with will be greatly expanded, as the number of novel functionalities which can be integrated on the chip can achieve.

This is a result of the miniaturization of mechanical features and fabrication and integration with electronics. However, there is another key characteristic of MEMS that goes beyond the simple reproduction on the small scale of macroscopic, well-known, devices. This is *scaling*.

Scaling is intended as the set of effects which arise and/or change in their intensity when the dimensional scale changes by one or more orders of magnitude.

Since MEMS have mechanical features 'sizing' from a few millimetres down to hundreds or even tens of micrometres, there might be a 10-fold to 1000-fold shrinking in linear dimensions, compared to structures in the centimetres size range. Since the physical phenomena involved in the operation of a device may depend, in general, by the system's linear dimensions to a certain power, the effect of such a dimensional scale change on the different parameters and laws governing the device will be weighted differently, and sometimes amazing results may arise.

As a typical example, one can report the relations among the linear dimensions, the surface and the volume of a given object, even for a simple cube for example, when its sides are isotropically scaled down by a factor of 10. The total surface of the objects then scales by a factor of 100, while its volume is 1000-fold smaller. This means, for example, that such an object will be 1000 times lighter but will experience only 1/100 of the friction when moving in a given medium.

If the object is a battery, its stored energy, which is proportional to volume, will be reduced 1000 times, while if the object is a source of radiating power, its energy transfer capability will be only 100 times smaller.

These examples may help to make clear, here, what scaling means and which consequence it has on the operation of devices. The relative influence of parameters and effects governing the operation of a given device changes when the linear dimensions of such devices are scaled down (or up). Phenomena which are commonly negligible to the macroscale may become important, or even dominant, at the microscale, and vice versa. Such changes in the relative importance of these phenomena may either favour or hamper the operation of a given miniaturized device, and this is very important to understand before facing the realization of a new device.

In all of those cases where scaling leads to advantages in the device's operation, this represents an added value for MEMS which goes beyond the simple fact of miniaturizing a macroscopic object.

There is another, maybe more important, key characteristic of MEMS. This derives from the phenomenon of 'change-of-weight', which follows a 'change-of-scale'. MEMS operate at a scale, the microscale, where the governing physical phenomena are the same of the 'macroworld', but in many cases they intervene with different weights with respect to the more common 'macroscale'. The different relative importance of physical effects and phenomena which can be experienced at the microscale allows for the conception of novel devices, which exploit different operating principles and perform better than their macroscale counterparts. In other words, MEMS may represent the enabling technology to a wider arena of potential applications which were inefficient, unfeasible or even unconceivable at the macroscale.

Such a type of regime of innovative applications and technology, which could be considered a significant portion of the 'Plenty of room at the bottom', envisioned by Richard Feynman in his historical lecture on miniaturization (Feynman, 1959), has not yet been fully explored and although many research efforts have been devoted to it.

The deep comprehension of scaling mechanisms and laws is of crucial importance for the effective development of these novel technologies and applications, since a critical and careful analysis of scaling leads to better optimized solutions, exploiting advantageous operating principles and working more efficiently. This means that scaling is an important design parameter which helps the designer in the choice of the best sensing/driving method for a given application at a given scale.

Thanks to miniaturization, integration with electronics and the phenomenon's change-of-scale, MEMS represents a technology to realize smaller, better and 'smarter' objects. In addition, scaling laws are the tool to apply such technologies in an efficient way.

A further 'shrinking' of both electronic and electromechanical devices will 'enter them' into new dimensional scales, i.e. the mesoscale and the nanoscale. Some quantummechanical effects become observable and therefore the device models must take these into account. In the case of electronic devices, some of these quantum mechanical effects may represent fundamental limits for further, future MOSFET scaling, while other effects provide the basic principles for new generations of electronic devices. In the case of electromechanical systems, this scaling step will provide higher sensitivities to the alteration of the system's physical properties, suitable instruments for the study of the new phenomena that characterize such a mesoscale and that represent the key for the comprehension of phenomena at the nanoscale (atomic or molecular scale).

In this chapter, the scaling paradigms for electronics and electromechanical devices, starting from the microscale and going down to the atomic and molecular scale, are analysed and compared.

1.2 EXAMPLES OF DIMENSIONAL SCALING POTENTIALS

In this section, some scaling effects and laws will be examined by taking into account some elementary microdevice structures.

1.2.1 Scaling effects on a cantilever beam

Consider a cantilever beam made of a given material (uniform and isotropic), having dimensions L, w and t (length, width and thickness, respectively). Let the cantilever have L along the x-axis, w along the y-axis and t along the z-axis.

Given ρ as the density of the material, the mass of the cantilever is:

$$M = \rho Lwt \qquad (1.1)$$

The elastic constant along the z-direction can be expressed as:

$$K_z = 12\frac{YI}{L^3} \qquad (1.2)$$

where Y is the material's Young modulus and I is the cross-sectional momentum of inertia, which is proportional to wt^3.

Given l as the generic linear dimension, M scales as l^3, and K_z scales as l, or linearly with l. Let's assume the notation $M = [l^3]$ and $K_z = [l]$

to indicate that M and K_z scale with the third and the first power of the linear dimension, respectively.

For example, if one scales the linear dimensions of a cantilever by a factor of 10, isomorphically, say $l' = 0.1l$, then the corresponding scaled mass and elastic constant are, respectively: $M' = 0.001M$ and $K_z' = 0.1K_z$. This means that the 10 times linearly scaled cantilever is 1000 times lighter, but only 10 times less stiff than its non-scaled counterpart; therefore, the scaled version has an improved mechanical robustness.

A suitable excitation can drive the cantilever to vibrate at its own resonance frequency ω, which is given by:

$$\omega = \sqrt{\frac{K}{M}} \tag{1.3}$$

Therefore, it can be easily seen that $\omega = l^{-1}$, which gives for the linearly scaled cantilever, $\omega' = 10\omega$.

The vibrating cantilever may be used as a mass sensor by measuring its resonance frequency shift with respect to a reference, or 'unloaded', value ω_0, due to a change in the mass:

$$M = M_0 + m + M_0\left(1 + \frac{m}{M_0}\right) \tag{1.4}$$

where M is the total mass, M_0 is the 'unloaded' mass and m is the mass change. Thus, the resonance frequency becomes:

$$\omega = \frac{\omega_0}{\sqrt{1 + \frac{m}{M_0}}} \tag{1.5}$$

The sensitivity of the resonance frequency to mass change can be expressed as the derivative of ω with respect to m:

$$S = \frac{d\omega}{dm} = -\frac{1}{2}\frac{\omega_0/M_0}{\sqrt{(1 + m/M_0)^3}} \tag{1.6}$$

where $S = l^{-4}$, which means that $S' = 10\,000S$, i.e. a linear scale factor of 10 in the dimensions of a cantilever beam leads to a 10 000-fold improvement in sensitivity of resonance frequency to mass change, which also means that smaller cantilever beams can potentially detect even smaller masses or mass changes.

The application of a force F to the cantilever tip, along the z-axis, will displace the tip by an amount δ, which can be thought of as a fraction of the cantilever's length L. F and δ are related by Hooke's law:

$$F = K_z \delta \qquad (1.7)$$

Then, the force required to achieve a given displacement δ scales as $F = [l^2]$ and thus $F' = 0.01F$. The same relative displacement can be achieved on a 10-fold scaled cantilever with a 100 times smaller force.

The amount of work performed (energy spent) to displace the cantilever tip is:

$$W = F\delta \qquad (1.8)$$

and this scales as $W = l^3$, which means that the energy required (actuation energy) to achieve a given relative tip displacement in a 10-fold smaller cantilever is 1000 times less than the non-scaled one.

Another example of scaling effects will be analysed by taking into account the operation of a cantilever-beam-based device as an inertial sensor. Since a cantilever beam is a suspended mass, it could, in principle, be operated as an inertial sensor to measure, for example, acceleration. Such a system can be modelled from a mechanical point of view, as a mass-spring-damper system, as represented in Figure 1.1. If the system experiences an acceleration, the mass is subjected to a force proportional to the acceleration, which is contrasted by the spring elastic reaction and the damping given by the surrounding medium. At equilibrium, the acceleration produces a net displacement of the mass position (i.e. the cantilever tip). Newton's 2nd Law gives:

$$M\ddot{x} + D\dot{x} + Kx = -Ma \qquad (1.9)$$

By supposing that the acceleration a has a sinusoidal dependence on time, the previous equation can be solved in the frequency (or $j\omega$) domain:

$$-\omega^2 Mx + j\omega Dx + Kx = -Ma \qquad (1.10)$$

The acceleration can be read by measurement of the displacement of the proof mass (cantilever beam) or the mechanical stress in the spring

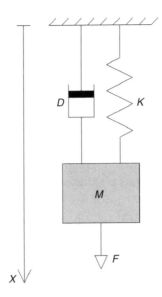

Figure 1.1 Simplified mechanical modelling of an inertial sensor as a mass-spring-damper system

(cantilever body), which is proportional to the force (acceleration); therefore, the frequency response of the system in terms of mass (tip) displacement vs. acceleration is:

$$x = \frac{-M/K}{1 + j\omega\dfrac{D}{K} - \omega^2\dfrac{M}{K}} \tag{1.11}$$

and the resonance frequency of the system is:

$$\omega_0 = \sqrt{\frac{K}{M}} \tag{1.12}$$

If the acceleration a is constant or slowly varying with time, the previous equation for the system's frequency response can be approximated by its DC value – thus:

$$x = -\frac{M}{K}a \tag{1.13}$$

The sensitivity of such an inertial sensor in terms of mass (tip) displacement vs. acceleration is, therefore:

$$S = \frac{dx}{da} = -\frac{M}{K} \tag{1.14}$$

It has been shown that the spring constant for such a cantilever beam scales linearly with linear dimension and thus if the simple inertial sensor described here is isomorphically scaled down by a factor of 10, its sensitivity to acceleration has a 100-fold reduction, i.e. the same acceleration produces a 100 times smaller displacement or, with the same achievable displacement, one can measure accelerations 100 times higher; thus, scaled structures are more suitable for the measurement of higher accelerations.

The example described here highlights another advantage of scaling on a simple structure such as a cantilever beam. Of course, the overall performance of an inertial sensor, as any kind of sensor, will depend on the adopted readout strategy and circuits. However, this is of no concern for this section and, moreover, it does not affect the generality of this discussion.

Further discussions on mechanical sensors will be addressed in Chapter 5, along with some other examples and cases study.

1.2.2   Scaling of electrostatic actuators

Other basic scaling effects can be examined by taking into account different structures and/or principles. The case of an electrostatic actuator based on a parallel-plate capacitor configuration, as shown in Figure 1.2, will be considered here.

The electrostatic energy W_e stored in such a capacitor, when it is charged to a voltage difference V between the two plates is:

$$W_e = \frac{1}{2}CV^2 = \frac{1}{2}\varepsilon_0\frac{wv}{d}V^2 \qquad (1.15)$$

For a given structure and dielectric material, the stored energy is 'upper-limited' by the dielectric breakdown voltage V_b, and therefore the

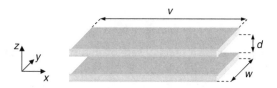

Figure 1.2   Parallel-plate capacitor

maximum energy is:

$$W_{e,m} = \frac{1}{2} \varepsilon_0 \frac{wv}{d} V_b^2 \qquad (1.16)$$

In the case of a parallel-plate capacitor structure, the dielectric break-down voltage can be written in terms of the breakdown electric field for the dielectric and its thickness, $V_b = E_b d$, and then is:

$$W_{e,m} = \frac{1}{2} \varepsilon_0 wvd E_b^2 \qquad (1.17)$$

Since the permittivity and breakdown field are material properties, it follows that the electrostatic energy stored in a parallel-plate capacitor scales as its volume.

The analysis of scaling for electrostatic actuators allows the observation of a first phenomenon due to change of scale. In the previous discussion about the maximum voltage a dielectric layer can sustain, it has been supposed that the breakdown electric field does not depend on dimensions; this leads to a breakdown voltage which scales linearly with the dielectric thickness. The thinner the dielectric, the lower is the voltage it can safely sustain, and therefore the smaller is the energy it can store. As a consequence of such a scaling effect, miniaturization of electrostatic devices, and especially actuators, appears to be inefficient.

If the dielectric is a gaseous one, as supposed in the previous analysis, the conclusion commonly holds at the macroscale, while a significant deviation from such behaviour can be observed when the dielectric thickness approaches the micron range. The breakdown of dielectrics is an 'avalanche effect'; thus, it appears at field values as smaller as the higher is the number of potential charge carriers that can be generated and this contributes to the effect. As the dielectric thickness is reduced, the total number of atoms or molecules contained in the gas volume decreases, and then the probability of collisions is reduced, which translates in an effective increase of the breakdown threshold. This is known as the *Paschen effect* and represents one of the phenomena that become more important in a change of scale, and in particular it allows realizing electrostatic microdevices more efficiently than one would predict on a first view. For example, a breakdown electric field of the order of 10^8 V/m has been reported for small gaps (Bart *et al.*, 1988), while the highest breakdown field measured in vacuum is reported to be 3×10^8 V/m (Madou, 2002), which are well above the 3×10^6 V/m observed at the macroscale.

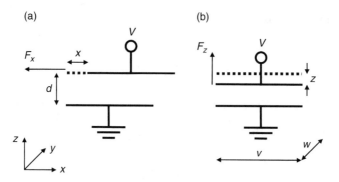

Figure 1.3 Actuation forces on parallel plate-capacitors

To estimate the maximum force an electrostatic actuator of the type shown in Figure 1.2, the two cases illustrated in Figure 1.3 will be considered here.

In the two parts of Figure 1.3, the parallel-plate capacitor is considered to be initially charged at a voltage V. If the upper plate is displaced by a small amount along the x-direction, as in Figure 1.3(a), an electrostatic force F_x that tends to realign the plates arises:

$$F_x = -\frac{\partial W_e}{\partial x} = \frac{1}{2}\varepsilon_0 E^2 v d \qquad (1.18)$$

In the same way, if the upper-plate is displaced by a small amount along the z-axis, the electrostatic force F_z is:

$$F_z = -\frac{\partial W_e}{\partial z} = \frac{1}{2}\varepsilon_0 E^2 v w \qquad (1.19)$$

In both cases, the electrostatic forces is proportional to the applied electric field E and the product between the dimensions transversal to the displacement direction; thus, the electrostatic force scales as the square of the linear dimension. Since the electric field is 'upperlimited' by the breakdown field value, E_b, then thanks to the Paschen effect, a gain in terms of force is achievable in electrostatic microactuators, compared to that estimated with macroscale knowledge, and also compared to the inertial forces which scale as the object's volume.

A deeper analysis will show, beyond the equations written above, for electrostatic microactuators of the types shown in Figure 1.3. that (a) performs better than (b), because in the latter the force depends on the plate displacement in common operating conditions; moreover, in such a kind of motion the plate experience 'squeeze-film damping', which leads to higher losses, and then lower quality factors for resonant mechanical devices. In laterally driven electrostatic actuators, instead, forces are independent on displacement to a larger degree; they allow larger displacements and experience the less-dissipative Couette damping (Madou, 2002).

From the previous examples it should be clear that scaling analysis does not barely consist on the extrapolation of the order of magnitude change for a given phenomenon consequential to a dimensional change of scale, but also, and even especially, on a careful analysis of the models and laws commonly adopted, and valid at the macroscale, aimed to find if other effects usually neglected become more important and/or new, unpredicted ones arise, and to determine whether such effects may represent a potential gain in performance for the microscale device.

1.2.3 Scaling of thermal actuators

Thermal actuators, together with electrostatic ones, are extensively used in MEMS and many examples can be found in the literature. The operation of thermal actuators is based on the thermal expansion of solids or fluids when they experience a temperature change or gradient. An exception is represented by shape-memory-alloy (SMA) actuators, which are based on crystalline phase transitions of particular materials. Thanks to their simple operating principles, thermal actuators can be realized with simple structures and in almost any technology.

Thermal actuators will be addressed more deeply in the following chapters of this book; however, a simple structure is considered here to provide an insight to the scaling of such devices. The simplest thermal actuator that can be imagined is a given volume V of a material having finite coefficients of thermal exchange and thermal capacity; when the temperature changes from T to $T + \Delta T$, the volume expands of an amount ΔV. If this expansion is suitably exploited, for example, by means of leverages and springs, it can perform physical work. For the purpose of simplicity, a cantilever beam having one dimension higher than the others, i.e. $L \gg w, t$, the thermal expansion can be considered as a thermal elongation ΔL:

$$\Delta L = L\alpha\Delta T \qquad (1.20)$$

where α is the coefficient of thermal expansion (CTE). The thermal expansion is counteracted by the elastic forces; the potential elastic energy is then:

$$\Delta W_K = -\frac{1}{2}K\Delta L^2 \qquad (1.21)$$

The force exerted by the thermal actuator in the direction of the thermal expansion is:

$$F = -\frac{\partial W_K}{\partial L} = K(\alpha\Delta T)^2 L \qquad (1.22)$$

The force scales as the square of the linear dimension, since the elastic constant K scales linearly.

In this example, it has been shown that thermal and electrostatic actuators scale with the same law in terms of forces, and then it could appear that the two kinds of actuators are equivalent to some extent. However, the comparison can be better made in terms of the actuator's stored energy density. In fact, as it has been shown, the force an actuator can exert is related to the amount of energy it stores. The energy density takes into account the storing efficiency of a given actuation scheme. Moreover, other performance figures, such as speed of operation, number of actuation cycles, fatigue, robustness to shock, etc., are usually of interest and therefore they should be taken into account when comparing the scaling of different actuation schemes.

The energy density for an electrostatic actuator is limited by the breakdown electric field, E_b and thus its theoretical maximum value amounts to:

$$w_{e,m} = \frac{1}{2}\varepsilon_0 E_b{}^2 \qquad (1.23)$$

If the dielectric is air or vacuum, the breakdown field can be as high as 3×10^8 V/m, and then the maximum energy density for electrostatic actuators is $w_{e,m} = 3.98 \times 10^5$ J/m^3.

Thermal actuators, thanks to the finite thermal capacity c, store energy in the form of thermal energy W_{th}, which in the stationary regime operation can be written as:

$$W_{th} = c\rho V\Delta T \qquad (1.24)$$

It is then straightforward that the thermal energy density stored in the actuator is:

$$w_{th} = c\rho\Delta T \qquad (1.25)$$

For a given material, the thermal energy density is limited by its melting temperature, or a temperature above which its physical properties change significantly. If the thermal actuator is a cantilever beam made of silicon, for example, whose density is $2330\,Kg/m^3$, thermal capacity $c = 711.75\,J\,K^{-1}\,Kg^{-1}$ and a melting temperature $T_m = 1687\,K$, the maximum theoretical stored thermal energy density is $2.8 \times 10^9\,J/m^3$. This performance estimation for thermal actuators, however, is too optimistic; in fact, the maximum allowable temperature is determined not by the melting of silicon, but by the lower melting-point materials used in most IC technologies, such as the aluminium-based alloys used for metallization, which melt at around $700\,K$. In the hypothesis that the maximum operating temperature is maintained well below half the melting point value, say $350\,K$, the stored energy density for such a thermal actuator amounts to $5.8 \times 10^8\,J/m^3$, which is still more than three orders of magnitude higher than those estimated for electrostatic actuators, even if the scaling-advantageous Paschen effect is taken into account.

The previous result does not mean that thermal actuators are better, or scale better, than electrostatic ones in an absolute sense. It means, instead, that thermal actuators allow higher energy densities, and then higher forces or displacements to be performed, but this has a cost in terms of much higher power supply requirements for thermal actuators when compared to electrostatic ones. Moreover, the operation of thermal actuators is limited in speed by the characteristic times of thermal exchange phenomena. In fact, to perform actuation, the device has to be cycled between two limit temperatures; thus, heat has to be effectively provided to, and removed from the device body, and these transfers are determined by the thermal time constant τ_{th}, which is defined as follows (Baglio, 2002):

$$\tau_{th} = \frac{\rho c V}{h A} \qquad (1.26)$$

where h is the heat exchange coefficient, which depends on the heat exchange phenomena involved (conduction, convection or radiation), and A is the surface area through which heat is exchanged. From equation (1.26) it results that the thermal time constant scales linearly with the system dimensions and therefore it is expected that thermal actuators achieve gain in terms of speed from miniaturization.

1.3 MOTIVATION, FABRICATION AND SCALING OF MEMS

MEMS, or *microsystems*, represent a wide class of devices which can be realized in many technologies and processes, in many different materials, and which can exploit many different principles. However, throughout this work, the focus is toward such devices which can be realized with silicon IC-compatible technologies and materials only, because they can potentially exploit all of the advantages of these low-cost, large volume technologies, including miniaturization and integration with electronics.

In the previous section, the potential of scaling electromechanical devices in the millimetre down to the micrometre range, together with the potential advantages of such scaling, has been anticipated. This is basically the answer to the question 'why electromechanical systems should be scaled?'. The spectrum of novel, potential applications for a given device spreads as its features size are reduced one order of magnitude or more. Furthermore, its performance may take advantages from such scaling.

However, the previous discussion has only a speculative value until a reliable, cost-effective, large-scale and functional fabrication technique is developed for the miniaturized devices. In the field of MEMS, IC technologies, and especially silicon-based ones, represent the key to the implementation of all of the advantages characterizing a miniaturized device, the most effective approach to the realization of efficient, cheap and 'smart' microelectromechanical systems.

Microelectronics IC technologies are based on two-dimensional, or planar, pattern-transfer techniques, selective implantation, growth, deposition and etching of substances and materials on a semiconductor (silicon) substrate. The result of these processes is a set of planar structures implementing given circuit functions. A given mechanical function, instead, needs a certain degree of freedom in the space to be performed, usually resulting in the movement or deformation of parts of solid bodies (cantilevers, beams, plates). The standard IC fabrication techniques do not allow the realization of suspended or free-moving mechanical parts on the semiconductor substrate or any other functional material; thus, a set of additional process steps are needed in order to realize such mechanical parts, while still maintaining the full compatibility with the overall technology to preserve the integrity and functionality of the electronics. Many fabrication, or micromachining, processes have been introduced in the literature and adopted for the fabrication of prototype or commercial MEMS devices, but two of them are the most widely used, thanks to their simplicity, which allows minimizing the number of additional steps, and their good degree of compatibility with standard materials and processes. These techniques are known as *bulk* and *surface micromachining*, and they are comprehensively treated in many reference books (Madou, 2002; Kovacs, 1998; Gad-El-Hak, 2001).

Silicon IC technologies, together with compatible micromachining processes, respond to the question about 'how electromechanical systems can be scaled?'.

The third, most important, question, which is the support of the leading idea throughout this work, is 'how much a given electromechanical system can be conveniently scaled?'.

At the microscale, the physical phenomena are the same as for the macroscale, because the scale is still too large to observe quantum effects, but the relative weight of the phenomena may change. Therefore, some approximations that are valid at the macroscale may not be useful or accurate enough at the microscale; thus, proper model reviews are required. Furthermore, due to this change of relative influence of physical phenomena, effectiveness of sensing and actuation means may change at the microscale to a degree that some methods gain in terms of sensitivity, while other may become unpractical. So, the problem questioned here is relative to the degree of scaling that can be conveniently applied to a given sensing or actuation method or structure.

It is the basic idea and the aim of this work to develop an analysis and design methodology for MEMS which is based on the critical study of scaling laws and their implications, and which may represent the answer to this fundamental question.

Typical feature sizes of MEMS range from some millimetres down to micrometres; a further scaling step down to the sub-micrometre range would enter these devices into the size regime that Roukes (Roukes, 2001) calls the *mesoscale*, which ideally extends from 100 nm down to 1 nm.

The mesoscale represents an interesting transition regime since objects in this dimensional scale are not so small to be easily described in terms of quantum mechanics, although they are yet not big enough to be free of quantum effects, thus allowing observation of many interesting properties, as described in Rouke's work, for example. The arise of such quantum effects represents a breakthrough in the classical comprehension of physical phenomena and in scaling laws, some of which begin to fail at the mesoscale.

The interest toward nano-electromechanicalsystems (NEMS) is motivated by the possibility to sense smaller quantities (forces, displacements, masses, etc.) and also to generate and control them on such a smaller scale. Moreover, NEMS represent a suitable platform for the study of the phenomena characterizing the mesoscale.

However, it is beyond the scope of this work to address the quantum effects that arise at the mesoscale. However, some NEMS examples will be glimpsed here to estimate the potential of a further dimensional scaling downward towards the micrometre range, and to envision which are the ultimate limits to this scaling approach.

1.4   SCALING AS A METHODOLOGICAL APPROACH

The scaling of microelectronics technologies can be quantified in terms of a paradigm, usually consisting in a set of features and parameters which determine the main characteristics for a given technology. For example, in the case of CMOS technologies such a paradigm is represented by the channel length: once the channel length is given, the most important characteristics of the technology are determined. Moreover, almost all of the consequences of a scaling can be expressed and quantified in terms of such parameters or paradigms, at least until the quantum effects can be neglected.

The possible MEMS devices, applications and technologies are so widespread that it is impossible to find a scaling paradigm which works like the channel length for CMOS technologies. On the other hand, a general analysis of scaling would require a certain degree of abstraction, which would not be possible by considering all specific-case applications.

Together with the guidelines for a design and analysis methodology based on scaling laws, a scaling paradigm will be proposed in the following chapters of this work.

As the previous examples have shown, a careful analysis of the scaling laws and effects for a given device, not limited to a straightforward extrapolation from dimensional dependencies but which takes into account the change of scale for physical phenomena, allows the designer finding those effects, structures and parameters which can be optimized for the application of concern, thus leading to a very efficient design.

The systematic application of this idea is complicated by the wide spectrum of potential applications, devices and implementations, as for the previously introduced idea for a scaling paradigm. However, the analysis of many different devices and applications can contribute to find several guidelines, which hold generally enough, and to implement a sort of knowledge-base of specialized applications.

The idea of using scaling laws as a methodological approach to the analysis and design of MEMS will be proposed and developed throughout this work, giving the basics for the development of such a methodology and knowledge-base.

REFERENCES

S. Baglio, S. Castorina, L. Fortuna and N. Savalli (2002). Modeling and design of novel photo-thermo-mechanical microactuators, *Sensors Actuators A: Phys.*, **101**, 185–193.

S.F. Bart, T.A. Lobe, R.T. Howe, J.H. Lang and M.F. Schelecht (1988). Design considerations for micromachined electric actuators, *Sensors Actuators*, **14**, 269–292.

R.Feynman (1959). There's plenty of room at the bottom, available at [http://www.zyvex.com/nanotech/feynman.html].

K.J. Gabriel (1998). Microelectromechanical systems, *Proc. IEEE*, **86**, 1534–1535.

M. Gad-el-Hak (Ed.) (2001). *The MEMS Handbook*, CRC Press, Boca Raton, FL, USA.

G.T.A. Kovacs (1998). *Micromachined Transducers Sourcebook*, McGraw-Hill, New York, NY, USA.

M.J. Madou (2002). *Fundamentals of Microfabrication: the Science of Miniaturization*, 2nd Edition, M. Gad-el-Hak (Ed.), CRC Press, Boca Raton, FL, USA.

M. Roukes (2001). Plenty of room, indeed, *Sci. Am.*, 48–57.

2

Scaling of Microactuators – an Overview

An actuator basically converts a form of energy into mechanical energy, resulting in the motion of a moveable part based on electrostatic, magnetic, piezoelectric, thermal or the shape-memory effects. Moreover, the performance of a specific actuation strategy is usually qualified through evaluation and measurements of mechanical quantities such as displacement, linear or angular, force and moment or torque.

Among the above listed actuation strategies, some considerations will be made, in this chapter, about electrostatic actuators, magnetic and electromagnetic actuators, and particular attention will be dedicated to thermomechanical actuators. Each of these methods has its own specific advantage and they are actually the most used and integrated with photolithographic processes.

2.1 ELECTROSTATIC ACTUATORS

The main benefit of electrostatic actuation is its characteristic small power consumption. They are also largely used in MEMS due to other advantages such as sensitivity, fast response, precision, relatively easy fabrication, or integration with standard technologies (Rosa *et al.*, 1998). Recently, various electrostatic actuator designs, such as the parallel plate, laterally and rotary comb-drive and vertical comb-drive actuators, have been developed (Tang *et al.*, 1989; O'Shea *et al.*, 2005).

Scaling Issues and Design of MEMS   S. Baglio, S. Castorina and N. Savalli
© 2007 John Wiley & Sons, Ltd

They are all operated by charging two bodies, having several different geometries and extensions, with equal and opposite charges; then, the resulting electrostatic forces between the two bodies can be potentially generated about three orthogonal directions, depending on the adopted configurations. Both DC or AC currents can be applied for charging such electrodes.

Moreover, each of the named configurations has its own specific advantage. A serious problem, for example, that affects parallel-plate actuators, in which the electrostatic force is proportional to the inverse of the square of the gap between the plates, is the nonlinear phenomenon called 'pull-in', which seriously constrains the stable region at one-third the length of the gap (Chiou and Lin, 2005). The lateral comb-drive structure, in which pairs of interdigitated conductive fingers move one toward the other, has an advantage over parallel-plate actuators, since the electrostatic force is independent of the displacement of the actuator. Then, the positioning of the system can be accurately controlled. However, for such systems a high driving voltage is often required, a small displacement in DC driving mode is obtained and a large layout area is occupied.

2.1.1   Transverse combs modelling

Transverse interdigitated structures, referred to as devices operated along the direction normal to the finger axis, are often used as capacitive sensors and actuators.

It can be easily figured out how the displacement can be related to a variation in the capacitances between the fingers due to the variation in the air-gap thickness, named the 'gap'. If a structure like that reported in Figure 2.1 is considered, with the external finger's array being into the fixed part, the two capacitances C_1 and C_2 will vary oppositely as a consequence of the respective air-gap thickness variation with respect to the stator fingers.

The total amount of variation for each capacitance can be expressed by:

$$C_1(y) = N_r \left(\frac{\varepsilon_0 L t}{y_0 + y} + C_{fi} \right) \quad C_2(y) = N_r \left(\frac{\varepsilon_0 L t}{y_0 - y} + C_{fi} \right) \qquad (2.1)$$

whereas the total capacitance for $y = 0$ is:

$$C_{|y=0} = 2N_r \left(\frac{\varepsilon_0 L t}{y_0} \right) + C_f \qquad (2.2)$$

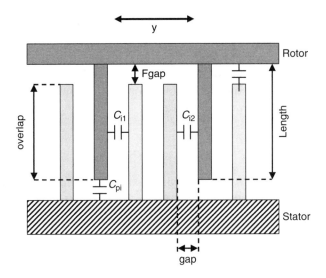

Figure 2.1 Schematic top view of a reference transverse comb

where L is the finger's overlap, y_0 is the initial gap between the fingers along the displacement direction, y denotes the displacement along the y-axis, N_r is the number of fingers of the moveable array (rotor), t is the finger's thickness, ε_0 is the dielectric constant, whereas C_f is assumed to be a 'parasitic' capacitance due to the fringe effects between adjacent electrodes, and $C_0 = 2N_r(\varepsilon_0 Lt/y_0)$.

The differential electrostatic force at each rotor finger, supposing to apply a voltage V_y at the rotor and a differential voltage $2V$ (with opposite sign at the stator's electrode couple) to the adjacent stator fingers, will be expressed by:

$$\Delta F = F_1 - F_2 \approx -\frac{1}{2}\frac{C_0}{y_0}\left[(V - V_y)^2 - (V + V_y)^2\right] \approx \frac{2C_0 VV_y}{y_0} \qquad (2.3)$$

where V is the polarization voltage. Then, a linear voltage-to-force function is obtained.

Finally, the device sensitivity can be expressed as:

$$S_c = \frac{\partial C}{\partial y} \approx \frac{C_0}{y_0} \qquad (2.4)$$

As an example, if $y_0 = 1\,\mu m$, $L = 50\,\mu m$, $t = 2\,\mu m$, $N_r = 50$ and a ratio $C_f/C_0 = 0.2$ are considered, then $C \approx 100\,fF$ and $S_c \approx 88.5\,fF/\mu m$. The device sensitivity benefits from the differential configuration, depends proportionally from the factor NLt and inversely from the gap y_0.

At the same time for such categories of devices, a contribution to the effective elastic constant of the structure must be taken into account due to the presence of the electrostatic force. It follows that an electrical elastic constant k_{el} can be defined as:

$$k_{el} = \frac{d}{dx}\Delta F = -\frac{2C_0 V V_y}{y_0^2} \qquad (2.5)$$

Then, the resonance frequency can also be significantly reduced for typical operating voltages:

$$\omega_r = \sqrt{\frac{k}{M}} = \sqrt{\frac{k_{mech} + k_{el}}{M}} = \omega_{r,mech}\sqrt{1 + \frac{k_{el}}{M}} \qquad (2.6)$$

Since the same topology can be exploited to realize electrostatic actuators, the displacement of the rotor can be evaluated by using the simulator (DIEES-MEMSLAB), which will be explained in Chapter 9, from the total electrostatic force and elastic constant through Hooke's law.

From the simulator point of view, it will be shown that pseudo-dynamic analysis is performed through computation of the steady-state response of the system to a selectable sinusoidal input.

2.1.2   Lateral combs modelling

Lateral interdigitated structures, referred to as devices operated along the direction of the longer dimension, are often used as capacitive sensors and actuators.

Here, referring to the symbols reported in Figure 2.2, the capacitance C can be expressed as:

$$C(x) = 2N_r\varepsilon_0\frac{(x_0 \pm x)t}{y_0} \qquad (2.7)$$

with the same meanings for the geometrical parameters as given in the previous section, y_0 is the transverse gap between the fingers (gap) that does not vary for this configuration, L_0 is the finger's length and $x_0 = L$ is the finger's overlap.

It follows that:

$$F_{el} = -\frac{1}{2}\left[\frac{C}{(x_0 \pm x)}\right]V^2 = -N_r\frac{\varepsilon_0 t V^2}{y_0} \qquad (2.8)$$

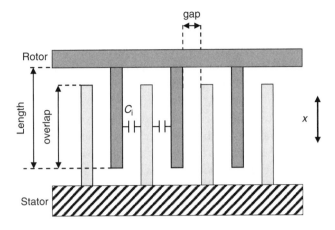

Figure 2.2 Schematic top view of a reference lateral comb device

Finally, the sensitivity can be derived from:

$$S_c = \frac{\partial C}{\partial x} = 2N_r \frac{\varepsilon_0 t}{y_0} \tag{2.9}$$

As an example if $y_0 = 1\,\mu m$, $L = 50\,\mu m$, $t = 2\,\mu m$ and $N_r = 50$ are considered, then $C \approx 9.2\,(fF)$ and $S_c \approx 0.4\,fF/\mu m$. Note that the electrical elastic constant is equal to zero in this case, whereas C is proportional to x and the sensitivity is generally poor.

As in the case of transverse combs, if the device is operated as an actuator the displacement in the x-direction can evaluated from Hooke's law.

If the static case is considered, the displacement of the structure used as an actuator can be derived from the static sensitivity of a second-order inertial system, since these devices are usually modelled through a mass-spring-damper configuration. Then, $x = F_{el}/k_{mech}$. Looking at equation (2.8), the electric elastic constant $k_{el} = 0$ for such a devices category.

Pseudo-dynamic analysis can be performed also in this case through the simulator described in Chapter 9, by imposing a suitable sinusoidal input and looking at steady-state solutions.

2.2 MAGNETIC TRANSDUCERS

Macroscale magnetic actuators have been widely used in applications ranging from motors to magnetic relays. Many design criteria and analysis methods have been developed to predict and to analyse their performance.

On the other hand, electromagnetic actuators are used extensively in the macroworld for different types of vertical, torsional and multiaxial actuation (Wagner and Benecke, 1991; Yanagisawa *et al.*, 1991; Holzer *et al.*, 1995).

In general, this technique can be used for microactuators by integration of coils and magnets with micromechanical elements. The efficiency of electromagnetic actuators is directly coupled to the material properties and the volume of magnets used (when they are coupled together).

Referring to the microworld, the most common electromagnetic actuation and sensing are based on the interaction between the electric current and an external magnetic field. The Lorentz force amplitude that corresponds to this interaction is defined by the product:

$$F_{L} = I_{f} L B \sin \vartheta \qquad (2.10)$$

where L is the length of a conducting wire and θ is the angle between the directions of I_{f} and B is the external magnetic field. By exploiting the conductor lines present in several MEMS technologies, rectangular or circular loops can be realized to carry the desired current I. Such loops, that can be obviously embedded inside a mobile structure, are then placed in an external magnetic field, considered, for instance, to be parallel to the plane of the loop, as shown in Figure 2.3. Simple considerations lead to the fact that there is no force acting on the arms

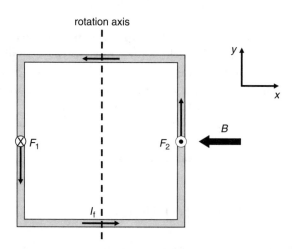

Figure 2.3   A loop carrying current immersed in a magnetic field B

of the loop parallel to the magnetic field. However, there will be two opposite forces acting with opposite sign on the other two arms, thus generating a moment which induces the loop to rotate about the y-axis.

The moment of this couple of forces is:

$$M = L^2 I_f B \tag{2.11}$$

where L is the arm's length.

Similar considerations can be made for multi-wire loops, both rectangular or circular. Figure 2.4 reports the layout and the microscope picture of a magnetic sensor which will be discussed in Chapter 4, based on this simple principle. This has been realized in CMOS (AMS 0.8 μm) technology and suitable 'Wet Bulk Micromachining' procedures based on TMAH.

Polysilicon-based strain gauges are embedded into the sustaining arms about which the sensor will rotate. The current path begins from an arm, patterned on the first metal layer of the adopted technology, turns about the central part and then ends up on the other arm through the metal 2 layer of the CMOS process, as illustrated in Figure 2.4(a). Then, the sensitivity of the device based on this structure increases proportionally to the active loop surface and number of turns, and inversely to the torsional elastic constant.

2.2.1 Magnetic actuators

Similar considerations can be made in defining 'magnetic' actuation or sensing. The role of the loop carrying the current can be played by a

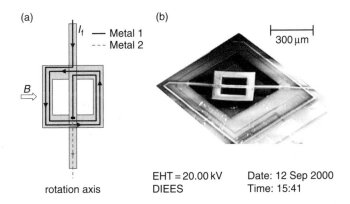

Figure 2.4 An example of a magnetic sensor based on the electromagnetic effect (Latorre *et al.*, 1998)

magnet, in which the magnetic field lines closing between the north and south pole of the magnetic dipole are equivalent to the current lines. When the magnet is suitably placed in an external magnetic field, a couple of counteracting forces will act on the magnet about a direction perpendicular to its plane, thus generating a moment that attempts to align the magnet to the magnetic field.

Such a moment can be expressed as:

$$m = MAL \qquad (2.12)$$

where M is the magnetization, A is the cross-sectional are a of the magnet and L is the magnet length. The magnetization can be then expressed for an isotropic material as:

$$M = B\frac{(\mu_r - 1)}{\mu_0 \mu_r} \qquad (2.13)$$

where μ_0 is the magnetic permeability of the free space and μ_r is the relative permeability of the magnet, defined as the ratio of its permeability to the permeability of the free space.

For anisotropic materials, the situation is a bit more complex because the relative permeability cannot be represented by a single value. An attraction force can be generated between a permanent magnet and a ferroelectric layer (which can be magnetized), as a means of magnetic transduction.

An interesting classification of magnetic actuators can also be carried out in two classes, as reported by Nami *et al.* (1996): those in which motion changes the gap separation (class 1) and those in which motion changes the gap overlap area (class 2) (Figure 2.5). Again, in both case it can be supposed that the magnetic energy is stored in the gap due to the large reluctance of the gap compared with the negligibly small reluctance of the magnetic core. However, in magnetic microactuators, the fabrication limitations on the achievable cross-sectional area of the magnetic core, as well as the finite core permeability, increase the core reluctance to the point that this assumption may no longer be valid. In this case, the magnetic energy is distributed in both the gap and the magnetic core, in which the energy distribution is in proportion to the reluctance of the gap (R_{gap}) and the reluctance of the core (R_{core}), respectively. Using an elementary structure of a magnetic actuator, it can be prooved that for the class 1 microactuators when the initial gap of the

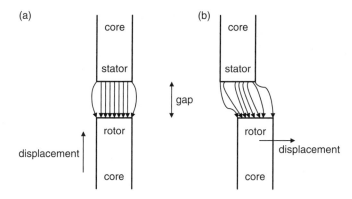

Figure 2.5 Schematic illustrations of the two different classes of magnetic microactuators: (a) class 1; (b) class 2 (Nami *et al.*, 1996). Reproduced by permission of IOP Publishing Ltd

actuator is fixed (e.g. determining the 'stroke' of the actuator), the generated magnetic force has a maximum value when the gap overlap area is designed such that the reluctance of the gap is equal to the reluctance of the magnetic core (i.e. $R_{gap} = R_{core}$). For class 2 microactuators, the initial overlap area of the actuator is fixed (determining the stroke); therefore, the generated magnetic force has a maximum value when the gap separation is designed such that the $R_{gap} = R_{core}$.

Therefore, in analysing magnetic microactuators, it is apparent that some of the assumptions which are perfectly adequate for macroscale actuators are unacceptable on the microscale. In Nami *et al.* (1996), it has been described which of these assumptions are not valid and why, and it has been shown that in certain cases the use of intuition and design criteria which are appropriate on the macroscale is inappropriate on the microscale.

In the conventional analysis of the macroscale magnetic actuator, most of the magnetic energy is, in fact, stored in the gap included in the magnetic circuit due to the large reluctance of the gap compared with the negligibly small reluctance of the core. However, in analysing magnetic microactuators, the assumption of gap reluctance dominance may not be acceptable on the microscale.

For magnetic microactuators, the reluctance of the magnetic core has a comparable value to or even exceeds that of the gap. In these reluctance-limited actuators, magnetic energy is distributed in both the gap and the magnetic core. By using a model structure of a magnetic actuator, it can be shown that the stored magnetic energy and the magnetic force generated in the gap have a maximum value when both reluctances of the gap and core are equal.

This condition would be an appropriate 'rule of thumb' in the design of magnetic microactuators for maximum force. As previously stated, in a class 1 actuator, the initial gap separation is determined by the desired actuator stroke. In this case, for maximum force, the gap area can be designed for reluctance equivalence. In a class 2 actuator, the initial gap area is fixed by the actuator stroke and it is the gap separation that can be adjusted according to the design criterion.

Since the result of reluctance equivalence holds for both types of actuators, it is necessary to examine in detail only one actuator type.

Before reporting magnetic circuit analysis on a given actuator, if a scaling boundary between the macroscale and the microscale could be defined, it would be convenient in determining whether conventional or reluctance-limited analyses are required. Unfortunately, no strict size-based rules exist to define the boundary between these analyses.

Although the selection of analysis cannot be determined based solely on geometrical size, it can be done based on relative core and gap reluctance values. For magnetic microactuators, core reluctances often increase to non-negligible values compared to gap reluctances due both to limitations on core thickness and achievable material permeability. For these reasons, magnetic microactuators commonly fall into the reluctance-limited or microscale analysis.

For the macroscale analysis, consider the class 1 magnetic actuator shown in Figure 2.6. The core is made of a magnetic material with a permeability μ and has a constant cross-sectional area, A_c. The gap, of

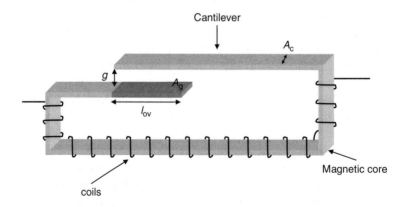

Figure 2.6 Schematic illustration of a magnetic actuator (Nami *et al.*, 1996). Reproduced by permission of IOP Publishing Ltd

area A_g and separation g, is shown in the structure. The total length of the core and the overlap length are denoted by l_c and l_{ov}, respectively.

In Figure 2.6, neglecting 'fringing', the magnetic reluctances of the gap (R_{gap}) and of the magnetic core (R_{core}) are as follows:

$$R_{gap}(x) = \frac{(g-x)}{\mu_0 A_g} \tag{2.14}$$

$$R_{core} = \frac{l_c}{\mu A_c} \tag{2.15}$$

where g is the initial gap separation before a motion takes place and x is the extent of motion normal to the surface of the gap electrode. It is assumed that the actuator motion does not alter the core reluctance. To consider the initial force on the actuator (i.e. the force prior to motion), x can be taken equal to zero. In this conventional case, the reluctance of the magnetic core is usually negligibly small compared with that of the gap; thus, it is assumed that most of the magnetic energy is stored in the gap. As the moving part is displaced due to the generated mechanical force at the gap, the inductance of the excitation coil varies as a function of position of the moving part due to the reluctance variation of the gap.

For a linear system, the energy and 'co-energy' are numerically equal. Therefore, the inductance (L), magnetic stored energy (W_m) and magnitude of the mechanical force in the x-direction (F_x) on the structure can be shown as follows (White and Woodson, 1959):

$$L = \frac{N^2}{R_{gap}x} \tag{2.16}$$

$$W_m = \frac{1}{2}i^2 L x \tag{2.17}$$

$$F_x = \frac{\partial W_m}{\partial x} = \frac{A_g}{2\mu_0}\left(\frac{iN\mu_0}{g-x}\right)^2 \tag{2.18}$$

where N is the number of coil turns and i is the excitation current. Note that it is assumed that the reluctance of the magnetic core is negligibly small. As equation (2.18) shows, the generated force increases as the actuation motion decreases the gap separation. Thus, in order to exert a maximum force in the actuator, the gap should be made as small as possible and the overlap area made as large as possible since the force increases monotonically as the gap separation shrinks and the gap area increases.

Then, in the conventional analysis, equation (2.18) shows that the smaller the gap separation and the larger the gap overlap area that is achieved, the larger the generated force, which can be considered as a design criterion for conventional magnetic actuators.

The corresponding design criterion for the microscale (i.e. when core reluctance can no longer be neglected) can be determined with respect to appropriate geometrical variables, to test whether or not the stored gap energy or the generated magnetic force has a maximum value at a given gap geometry.

In the class 1 microactuator, there are two ways to adjust the reluctance in the magnetic core (R_{core}) for a given permeability: the core cross-sectional area A_c and the core length l_c. The variation of the core cross-sectional area in a planar-type integrated inductive component is limited by the achievable thickness as well as the width of the core, as magnetic microdevices are usually implemented in a planar fashion. The core length l_c is mainly determined by the required number of coil turns necessary to attain a required magnetic flux density in the device. Consequently, the adjustment of the geometrical dimension to vary the magnetic core reluctance is usually limited.

However, the geometrical variables for the gap reluctance (R_{gap}) can be flexibly adjusted by varying the gap area A_g (or the overlapped length l_{ov}) and the initial gap g. In a class 1 microactuator, the initial gap is usually set by the desired range of actuation. In addition, for the microactuator shown in Figure 2.6, the maximum achievable gap separation is limited due to the limitation of the thickness of suitably deposited 'sacrificial' layers. Moreover, the gap area A_g (or the overlap length l_{ov}) can be easily adjusted by varying either the overlap gap width or length. Thus, it is found that the energy stored in the gap has a maximum value when $R_{gap} = R_{core}$.

Such a condition can serve as the criterion for maximum force in designing magnetic microactuators on the microscale. It is interesting to note that it is analogous to the maximum power-transfer condition in an electrical circuit to transfer a maximum power to a load (R_l) from a power source which includes an internal source resistance (R_s), where the condition is $R_l = R_s$. Thus, when the reluctance of the core cannot be neglected, careful sizing of the gap area and/or the initial gap separation are necessary to satisfy the maximum force design criterion of magnetic microactuators.

In conclusion, due to fabrication constraints, microactuators often operate in the reluctance-limited regime. In this regime, maximum energies and forces are found when the actuator is designed such that the gap and core reluctances are equal. For a class 1 microactuator, in which actuation occurs in the direction of the gap separation, a feasible design sequence would be to size the initial gap separation in accordance with the actuator stroke requirements, then size the gap area such that the reluctances of the gap and core are equal. For a class 2 microactuator, in which actuation changes the gap area but not the gap separation, a feasible design sequence would be to size the initial gap area in accordance with the actuator stroke requirements, then size the gap separation such that the reluctancess of gap and core are equal.

2.2.2 Ferromagnetic transducers[1]

Ferromagnetic materials have proven to be useful in a many macroscopic sensors (e.g. flux-gate magnetometers) and actuators (e.g. electromagnetic motors). Most ferromagnetic microsensors and microactuators are essentially scaled-down versions of macroscopic devices (e.g. electromagnetic motors (Guckel *et al.*, 1993; Ahn and Allen, 1992; Ahn *et al.*, 1993) and flux-gate magnetometers (Andò *et al.*, 2005; Ripka *et al.*, 2001) complete with ferromagnetic cores, copper windings, insulation, etc.). In addition, the magnetic recording head industry has done much to develop the low-cost micromachining processes needed to batch fabricate large numbers of magnetometers on a single substrate for data storage applications.

In nearly all cases, electrodeposition plays a key role in the formation of the ferromagnetic material due to its flexibility, capability, low-cost and quality.

The performances of magnetic and electrostatic devices, and some aspects related with scaling them, can be compared considering identical gap geometries, for instance, identical air gap and cross-sectional area (Judy *et al.*, 2000). In the electrostatic system (Figure 2.7(a)), it is assumed that no DC currents flow and that no electric field exists in the conductors connecting the voltage source to the plates at the gap. These assumptions are valid because the ratio of the conductivity of the conductor is significantly large. Such a large difference in conductivity allows electric fields to be tightly confined and controlled. In the magnetostatic system (Figure 2.7(b)), the ferromagnetic core material is assumed to have a length L and a finite relative magnetic permeability μ_r. Note that the ratio of the permeabilities of core materials does not

---

[1] This section is reproduced by kind permission of ECS.

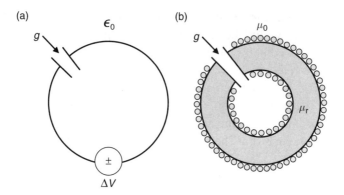

Figure 2.7 Structures with identical gap geometries for comparing (a) electrostatic and (b) magnetostatic forces (Judy *et al.*, 2000). Reproduced with kind permission of The Electro-chemical Society

guarantee that the magnetic field is tightly confined, meaning that there are no good magnetic insulators, except superconductive materials.

Both structures are assumed to be mechanically rigid and the 'fringing fields' near the air gap are ignored. Since the force F that an actuator can generate is equal to the negative spatial derivative of the energy U stored between the armature and the stator ($F = -\nabla U$), the stored energy density u is a useful figure of merit for comparing actuators of different sizes. If fringing fields are neglected, the electrostatic energy density u_e stored in the air gap between the electrodes is given by $u_e = \varepsilon_0 E^2/2$ with the permittivity of free space ε_0. Similarly, a figure-of-merit for magnetostatic microactuators is the magnetostatic-energy density u_m stored in the air gap between the magnetic pole tips, which is given by $u_m = B^2/2\mu_0$ with the permeability of free space μ_0 (supposing to ignore the magnetostatic energy stored in the pole pieces themselves – a good assumption if the relative permittivity of the pole material μ_r is much larger than the ratio of the length of the pole pieces to their gap, l/g).

The maximum electrostatic-energy density $u_{e(max)}$ stored in an actuator whose electrodes are separated by a distance g is equal to:

$$u_{e(max)} = \frac{\varepsilon_0 E_{max}^2}{2} = \frac{\varepsilon_0 V_{max}^2}{2g^2} \quad (2.19)$$

with the maximum voltage that can be sustained before breakdown, V_{max}. The maximum magnetostatic-energy density $u_{m(max)}$ stored in an actuator occurs for all practical purposes when the magnetic material becomes saturated and $B_{max} = M_s$, so that:

$$u_{m(max)} = \frac{B_{max}^2}{2\mu_0} = \frac{M_s^2}{2\mu_0} \quad (2.20)$$

Given that the speed of light can be expressed as:

$$c = \frac{1}{\sqrt{\varepsilon_0 \mu_0}} \qquad (2.21)$$

the ratio of the magnetostatic and electrostatic energy densities, $u_{m(max)}/u_{e(max)}$, becomes:

$$\frac{u_{m(max)}}{u_{e(max)}} = \frac{M_s^2}{\varepsilon_0 \mu_0 E_{max}^2} = \frac{c^2 M_s^2}{E_{max}^2} = \frac{c^2 M_s^2}{V_{max}^2} g^2 \qquad (2.22)$$

On a macroscale, with armature–stator separations greater than 1 mm, the maximum electric field is $\sim 3\,V/\mu m$ and equivalent to an energy density of $u_{e(max)} \approx 40\,J/m^3$. For a saturated iron system, the maximum flux density is equal to $M_s = 2.15T = 2.15Vs/m^2$, which corresponds to $u_{m(max)} \approx 1.84 \times 10^6\,J/m^3$. In these situations, their ratio is equal to 4.6×10^4, a good reason why macroscopic actuation is dominated by magnetostatic devices.

However, at a separation of several micrometers, Paschen discovered that the breakdown voltage reaches a minimum value (Paschen, 1889). When the electrode separation is reduced to less than several micrometres, the breakdown voltage and the breakdown field increases rapidly. A plot of the breakdown voltage as a function of the product of electrode separation and pressure, given in Figure 2.8, is known

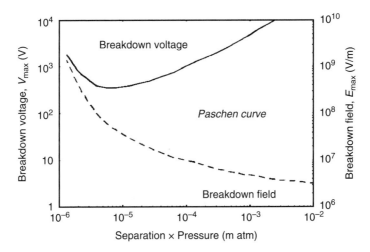

Figure 2.8 Plots of breakdown voltage and breakdown field as an function of electrode separation – known as Paschen's curve (Judy et al., 2000). Reproduced with kind permission of The Electro-chemical Society

as Paschen's curve (von Hippel, 1959). To the left of the minimum breakdown voltage, there are too few impacts to achieve a regenerative 'avalanche' breakdown unless the applied voltage increases by an extreme amount. To the right of the minimum breakdown voltage, the number of impacts is too large and much of the necessary energy is wasted in minor electronic excitation.

By using the breakdown voltage shown in Figure 2.8, the maximum electrostatic-energy density can be compared to the maximum magnetostatic energy density (Figure 2.9). In the ideal situation, where the stator and armature surfaces are perfectly smooth and fringing fields can be neglected, $u_{e(max)}$ exceeds $u_{m(max)}$ of a saturated iron system ($M_s = 2.15 \text{Vs/m}^2$) at stator-armature separations of less than $1.75\,\mu\text{m}$, although a voltage of over 1000 V is required for this to occur. For a saturated nickel system ($M_s = 0.6 \text{Vs/m}^2$), $u_{e(max)}$ exceeds $u_{m(max)}$ at separations less than $3\,\mu\text{m}$ and 500 V are required.

Furthermore, it is impractical to operate an electrostatic device very near to the Paschen-limit since tiny non-uniformities on the stator and armature surfaces (e.g. spikes, dimples and hillocks) will locally increase the electrical field and may cause premature electrostatic breakdown.

However, the energy density of an electrostatic system is limited by the amount of voltage and the size of the armature–stator gap that are considered practical to achieve.

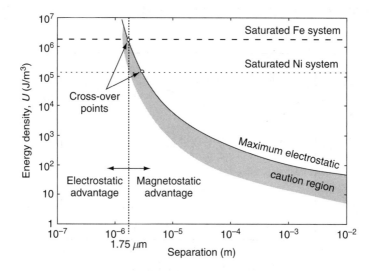

Figure 2.9 Maximum ideal energy densities for electrostatic and Ni or Fe magnetostatic systems (Judy *et al.*, 2000). Reproduced with kind permission of The Electrochemical Society

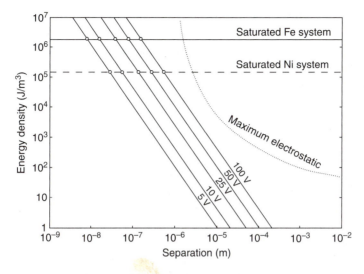

Figure 2.10 Energy densities for constant-voltage electrostatic systems and Ni or Fe magnetostatic systems (Judy *et al.*, 2000). Reproduced with kind permission of The Electro-chemical Society

Figure 2.10 shows a plot of the electrostatic energy density as a function of armature–stator separation for constant voltages of 100, 50, 25, 10 and 5 V. Equation (2.22) can be used to determine the armature–stator gap g needed to achieve $u_{ratio} = 1$ for a particular voltage electrostatic breakdown, $d = V/(cM_s)$.

For constant voltages of 100, 50, 25, 10 and 5 V, armature–stator separations of 155, 78, 39, 16 and 7.8 nm are needed to exceed the $u_{m(max)}$ of a saturated iron system. Thus, although electrostatic microactuators with armature–stator separations as small as 100 nm have been achieved by using novel device structures and fabrication techniques (Hirano *et al.*, 1992), these devices still need to be driven with more than 64 V to achieve $u_{ratio} = 1$. When working with an electrostatic microactuator with such small air gaps, care must be taken to prevent the electrodes from touching the device since such an electrical 'short' could fuse the device or otherwise render it inoperable. Magnetic systems, on the other hand, benefit from the fact that a magnetic stator and armature may come into contact without having a detrimental effect on the physical structure of the device. Magnetic microactuators have been produced that utilize on-chip magnetic fields to generate significant microactuation with stator–armature separations of a few micrometres.

Despite the substantial improvement in the energy densities (and hence the force) of electrostatic systems afforded by the effect illustrated by the Paschen curve (Figure 2.8), ferromagnetic MEMS are more powerful than electrostatic microactuators when micronscale gaps (or larger), typical voltages and typical currents are considered.

2.3 THERMAL ACTUATORS

Both electrostatic and thermal actuators, in principle, can be implemented by means of simple structures and with no need for special materials, which makes it possible to design and realize them in almost any IC technology. However, as already shown in Chapter 1, thermal actuators can be even simpler than electrostatic ones, and may develop large forces or displacement. On the other hand, electrostatic actuators have often a fast response and they can be easily integrated with standard or *ad-hoc* technologies.

Thanks to their characteristics, thermal actuators will be largely addressed here, and they will be taken into account as a 'test bench' for the scaling analysis.

Thermal actuators exploit the non-uniform heat distribution across a solid body due to non-homogeneous materials and/or nonsymmetric geometries. The non-uniform heat distribution induces stress and deformation in the actuator's structure, which ultimately results in displacement and/or the execution of physical work. This description of the operating principle is valid for almost any of the actuators of interest for this work, i.e. those which can be realized with technologies and materials compatible with conventional IC fabrication processes. Exceptions to this class are represented by fluid-based and shape-memory-alloy actuators, which operate with different phenomena.

Thermal actuators can be mainly: (a) simple beam (fixed free), (b) bent-beam (fixed–fixed), (c) two beam (fixed free) or (d) bylayer beam (fixed free). They are all based on a solid expansion effect under a temperature gradient and a particular mechanical arrangement which allows us to gain from suitable constraints of its parts to obtain larger sensitivities.

In the first case, a reference device composed of a cantilever of an homogeneous material experiences linear thermal expansion in the three Cartesian directions, proportionally to the material thermal expansion coefficient and temperature variation or gradient, as briefly discussed in Chapter 1 (Section 1.2.3). Usually, a preferential direction is chosen

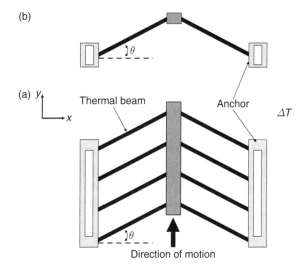

Figure 2.11 Structure and operation of (a) a basic single bent beam and (b) a bent-beam array actuator

for the designed bar and a predominant expansion in one direction, that of actuation, is considered, while the others can be supposed to be negligible.

An example of a bent-beam thermal actuator, discussed also by Gianchandani and Najafi (1996) and Que et al., 2001, is illustrated in Figure 2.11. When conceived as actuators, current passing through the suspended beams causes Joule heating, resulting in a displacement of the apex which is amplified by the bending angle of the beam. A linear motion in the y-direction is obtained due to the symmetry of the system, in both cases, and the combined bending of the bars. The two bars are, in fact, impeded to free expansion by the two supports; then, they can only bend one towards the other, so resulting in a deformation along the y-axis.

The governing equation of such an actuator derived by considering suitable boundary conditions and the second Castigliano's theorem – this will be explained in Chapter 9 – indicates that the deflection is linearly dependent on temperature increase and non-linearly dependent on the geometric features of the symmetric bent beams. The results reported by Que et al. (2001) indicate that the displacement produced by a bent beam is always larger than the axial deformation of a fixed-free beam for smaller inclination angles and larger lengths.

Although relatively large forces and substantial displacements can be generated, some important considerations can be made. For example, the motion is typically associated with the bending of a cantilever and is not rectilinear. Moreover, bimorphs are generally made for out-of-plane actuation (Riethmuller and Benecke, 1988). Although out-of-plane actuators can be used for in-plane displacements (Sun *et al.*, 1996; Okyar *et al.*, 1997), the in-plane forces are typically $< 10\,\mu N$. Arrayed out-of-plane actuators have been used as conveyors and positioners (Ataka *et al.*, 1993; Suh *et al.*, 1999; Ebefors *et al.*, 1999). Planar actuation (parallel to the device substrate) is particularly difficult to achieve since laying the bimorph materials side-by-side complicates the fabrication sequence, and does not necessarily produce a large rectilinear displacement.

Clearly, device performance can be altered by changing the geometry of the beam. In general, the peak displacement of a beam at a given temperature can be increased by making the beam longer or by reducing the bending angle. Larger displacements can be generated by 'cascading' the actuators. The numerical simulations reported in Figures 2.12 and 2.13, concerning planar bent-beams used as actuators and sensors, respectively. If the current provided or the gradient temperature are not varied, the sensitivity is enhanced by 'cascading' bent-beams. Amplification of motion is, in fact, obtained for actuators with respect to a single structure, as shown in Figure 2.12, as well as larger displacements obtained for sensors operating, for example, as shown in Figure 2.13. A capacitive output for such a thermal sensor is considered in this case. The sensitivity of such a structure, compared with that obtained from a single structure, is reported in Figure 2.14.

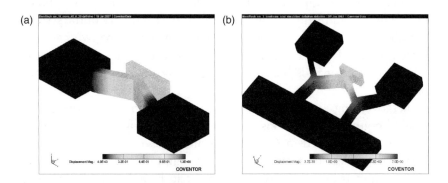

Figure 2.12 Deformed shapes obtained through the 'CoventorWare' simulator at 300 and 600 K, respectively, for (a) a single bent-beam device, composed of two arms and (b) an array of bent beams

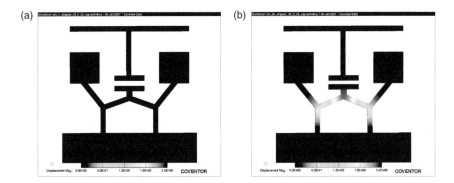

Figure 2.13 Top views of the deformed shapes for an array of bent beams obtained through the 'CoventorWare' simulator, at an ambient temperature of (a) 300 K and (b) 600 K, respectively. If the structure is composed of a conductive material and a second electrode is released, fronting the first, a capacitive sensing can be conceived

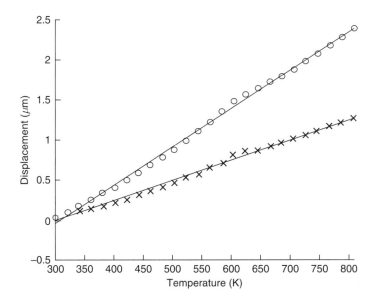

Figure 2.14 Comparison of sensitivities obtained from single (×) and arrayed (○) bent beams, corresponding to the designs of Figure 2.12

Numerical simulations have been made by using 'CoventorWare 2006', a powerful tool for MEMS designers. Each beam of the structure has a 40 µm length, a 20 µm thickness (to enhance the facing area for the conceived parallel-plate capacitor used for sensing the displacement of the device apex) and a $8\,\mu m$ width and $\theta = 20°$. The device apex is, in fact, a transverse bar with equal geometrical features.

2.3.1   Thermomechanical actuators

The analysis of operation and scaling for thermal actuators will be developed by taking into account the bilayer cantilever-beam structure shown in Figure 2.15. Despite its simplicity, this structure represents a suitable platform for the analysis, since it is widely applied and the procedure developed here holds even for different and more complex structures.

The cantilever beam shown in Figure 2.15 undergoes bending if exposed to a temperature change due to the difference in the coefficients of thermal expansion of the layers. In general, the cantilever can be made of more than two layers; however, the analysis for the simple two-layer structure still holds.

The tip deflection δ, due to a given temperature variation $\Delta\theta$, can be interpreted as the equilibrium position where the 'thermal driving force' F_{th}, bending the cantilever, is balanced by the elastic restoring force F_K. In other words, at equilibrium, the thermal energy W_{th} stored in the cantilever is equated by the potential elastic energy W_K, expressed as follows:

$$W_{th} = \rho c V \Delta\theta = \frac{1}{2}K_z\delta^2 = W_K \qquad (2.23)$$

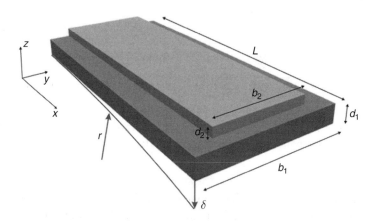

Figure 2.15   A bilayer cantilever-beam structure for a thermal actuator (Baglio *et al.*, 2002). Reproduced with kind permission from Elsevier

and then:

$$\delta = \sqrt{\frac{2\rho c V \Delta \theta}{K_z}} \qquad (2.24)$$

The thermal driving force F_{th} is, therefore:

$$F_{th} = -F_K = K_z \delta = \sqrt{2K_z \rho c V \Delta \theta} \qquad (2.25)$$

In the previous equation (2.25), the density and the specific heat are material properties which can be reasonably considered to be independent from the system's size to the extent of this treatment. The same holds for the temperature change. The elastic constant for a cantilever beam is given by the following equation (Gere and Timoshenko, 1997):

$$K_z = 12 \frac{YI}{L^3} \qquad (2.26)$$

where Y is the Young's modulus, which is a material property, and I is the momentum of inertia of the beam's transversal section and has units of length to the fourth power – therefore, the elastic constant scales proportionally to the system's linear dimension. From equation (2.25), it follows that the thermal driving force scales as the square of the system's size.

The energy density of a thermal actuator can be calculated from equation (2.23), by dividing for the volume V:

$$w_{th} = \frac{dW_{th}}{dV} = \rho c \Delta \theta \qquad (2.27)$$

By considering, for example, the structure and the average materials properties reported by Baglio *et al.* (2002), i.e. $c = 800\,J/Kg/K$, $\varrho = 2400\,Kg/m$ and $\Delta \theta = 100\,K$, it follows that the energy density amounts to $192 \times 10^6\,J/m$, higher than that achievable with electrostatic actuators. These results indicate that thermal actuators scale advantageously in terms of force and energy density. However, this analysis does not take into account the transient behaviour of the device, which is dominated by the heat-exchange phenomena. The worst case, i.e. the slowest heat-exchange effect, happens when the device has to dissipate all of the stored thermal energy through free thermal convection; in such a case, transient behaviour is regulated by the thermal time constant τ_{th}, which is defined as:

$$\tau_{th} = \frac{\rho c V}{hA} \qquad (2.28)$$

where h is the coefficient of convection and A is the area of the exchanging surfaces. By considering only the dependences of V and A from the device linear dimensions, it results that the thermal time constant linearly scales as the device sizes shrinks. This is undoubtedly an advantage for miniaturized thermal actuators with respect to their 'macro' counterpart. However, as it will be shown, heat exchange phenomena, and especially those where free convection in air is involved, are extremely 'slow' and therefore they heavily limit the maximum operating frequency of even microactuators, thus relegating them to static or low-frequency applications.

Thermal actuators can be easily realized in almost any semiconductor technology where a micromachining process can be applied. In fact, the basic structure of a thermal actuator is composed of a heater or, more generally, a source of heat, and a mechanical structure with some asymmetry in its geometry and/or in its thermal properties. Thermal actuation with solids is based on the asymmetric 'reaction' of this kind of devices to a temperature change.

A bilayer cantilever, as shown in Figure 2.15, is composed of two layers of materials with different coefficients of thermal expansion. As the temperature increases with respect to a reference value, a bending of the structure occurs due to the different expansions of the materials. The displacement of the cantilever free end in the direction perpendicular to the layer surfaces, with respect to an equilibrium position, is expressed by equation (2.24) or, as reported by Webster (1998) with a higher degree of detail:

$$\delta = \frac{3\,(1+m)^2}{3\,(1+m)^2 + (1+mn)\left(m^2+\frac{1}{mn}\right)} \frac{L^2}{d}\,(\alpha_2 - \alpha_1)\,\Delta\theta \qquad (2.29)$$

where α_1 and α_2 are the thermal expansion coefficients of the low-expansive material and of the high-expansive material element, respectively. In addition, $n = Y_1/Y_2$, with Y_1 and Y_2 their respective Young's moduli, $m = d_1/d_2$, with d_1 and d_2 their respective thicknesses, and $d = d_1 + d_2$, the total thickness of the cantilever.

The analysis of the heat-exchange and thermal phenomena involved in the operation of such a device is necessary to correlate the displacement or the force produced by the actuator with the power density provided by the heater. Such an analysis will provide the design equations and will allow estimating the system efficiency.

From the thermal-analysis point of view, the structure has a portion of its surface exposed to a pulsed heat flux (if the heater is supposed to be realized on the surface of the cantilever), while the rest of the structure exchanges heat with the surrounding air by convection. This clearly is a worst-case analysis since the device can, in general, exchange and dissipate heat through several mechanisms, for example, by conduction through the substrate. The thermal energy stored in the device can be converted into useful work.

The maximum operating temperature for the device is limited by the lower-phase change limit of the materials used to build the system. Phase change is intended as a set of phenomena, triggered by temperature, which cause an irreversible change of the geometric and/or mechanical properties of one or more materials. Examples include the melting of a material, a permanent deformation, a transition from an elastic to a plastic phase, etc.

The thermal transient behaviour of the device during the heating and cooling phases will be analysed here. The thermal transient analysis of the system can be performed by applying the so-called 'lumped capacitance method' (Incoprera and De Witt, 2001). Such a method requires the assumption of a uniform temperature distribution within the solid. The temperature gradient within the solid can be neglected if the *Biot number (Bi)*, defined as the ratio between the thermal resistance to conduction and the thermal resistance to convection, is much smaller than unity, expressed as follows:

$$Bi = \frac{R_{\text{cond}}}{R_{\text{conv}}} = \frac{hL_{\text{c}}}{k} \qquad (2.30)$$

where h is the coefficient of convection with air, k is the thermal conductivity and L_{c} is a characteristic length, which in this case is defined as the ratio between the volume of the solid and its surface area.

In the following analysis, the assumption of uniform temperature distribution will be made. It will be verified through the Biot number, successively.

The heat exchanges involved in the actuation scheme are shown in Figure 2.16. The energy balance of the system gives:

$$q_{\text{h}}A_{\text{h}} - h\,(\theta - \theta_{\infty})\,A_c = \rho c V \frac{\mathrm{d}\theta}{\mathrm{d}t} \qquad (2.31)$$

The first term on the left-hand side of equation (2.31) represents the incoming heat flux (q_{h} is the heat power density and A_{h} is the heated surface) while the second term represents the outgoing heat due to thermal convection with air (h is the coefficient of convection, θ is the

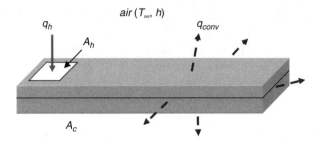

Figure 2.16   Schematic of the 'thermal problem' of the actuator (Baglio *et al.*, 2002). Reproduced with kind permission from Elsevier

temperature of the solid at time t and θ_∞ is the temperature of the 'unperturbed' air).

The cantilever is inhomogeneous due to its bilayer structure; however, for the sake of simplicity it also could be thought of as being made of a 'thermally equivalent' homogeneous material. In equation (2.31), the term on the right-hand side represents the stored energy (ϱ is the equivalent density, V the volume and c the thermal capacitance of the equivalent solid material).

By assuming that the incoming heat flux is constant (at least in finite time intervals, like in the case of pulsed source) and solving equation (2.31) with the initial condition $\theta(0) = \theta_0$, gives:

$$\frac{\theta - \theta_\infty}{\theta_0 - \theta_\infty} = \exp\left(-\frac{t}{\tau_{th}}\right) + \frac{q_h A_h}{h A_c (\theta_0 - \theta_\infty)}\left[1 - \exp\left(-\frac{t}{\tau_{th}}\right)\right] \qquad (2.32)$$

where the thermal time constant, τ_{th}, as defined in equation (2.28), is:

$$\tau_{th} = \frac{\rho c V}{h A} = R_{conv} C_{th} \qquad (2.33)$$

in which R_{conv} is the resistance to convection and C_{th} is the 'lumped' thermal capacitance.

The right-hand side of equation (2.32) contains a term which represents the cooling or heat loss transient, and a term which represent the heating transient, respectively.

To further develop the thermal analysis introduced in the previous section, and also to provide some numerical examples, a possible realization of a device prototype is proposed and addressed here. For this purpose, a standard CMOS technology will be taken into account, together with a bulk anisotropic etching process of the silicon substrate

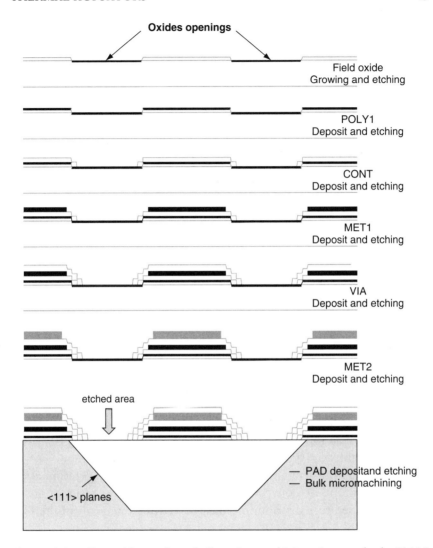

Figure 2.17 Front-side surface bulk micromachining in standard CMOS technology (Baglio *et al.*, 2002). Reproduced with kind permission from Elsevier

(Baglio *et al.*, 1999), which allows the realization of suspended structures made by layered oxides, polysilicon, metals and passivation, such as those shown in Figure 2.17.

The adoption of standard IC technologies, coupled with compatible post-processing, is a largely addressed strategy, in particular, for the wide interest it arises in the industrial world. The deposited films of a typical standard CMOS technology, with some of their geometrical and

Table 2.1 Thicknesses and physical properties of standard CMOS technology layers (Baglio *et al.*, 2002). Reproduced with kind permission from Elsevier

Layer	Name	d (nm)	$\rho(\mathrm{Kg\,m^{-3}})$	Y (Pa)	$\alpha(\mathrm{K^{-1}})$	$k(\mathrm{Wm^{-1}\,K^{-1}})$
Field Ox	DIFF	400	2.5×10^3	7.3×10^{10}	0.4×10^{-6}	1.10
Poly 1	POLY1	300	2.3×10^3	1.9×10^{11}	2.5×10^{-6}	157.00
Poly 12Ox	CAPOX	40	2.5×10^3	7.3×10^{10}	0.4×10^{-6}	1.10
Poly 2	POLY2	300	2.3×10^3	1.9×10^{11}	2.5×10^{-6}	157.00
Cont Ox	CONT	730	2.5×10^3	7.3×10^{10}	0.4×10^{-6}	1.10
Metal 1	MET1	600	2.7×10^3	7.0×10^{10}	23.0×10^{-6}	237.00
Via Ox	VIA	800	2.5×10^3	7.3×10^{10}	0.4×10^{-6}	1.10
Metal 2	MET2	1050	2.7×10^3	7.0×10^{10}	23.0×10^{-6}	237.00
Passivation	PAD	985	3.1×10^3	3.8×10^{11}	—	—

physical properties (Gad-el-Hak, 2001), are reported, in their process sequence order, in Table 2.1.

In the example proposed here, the bilayer cantilever structure is obtained by the superimposing of two oxide layers (DIFF and CONT) and two metal layers (MET1 and MET2).

Once the technology and a structure's hypothesis have been introduced, the detailed analysis design of a prototype can be addressed here. A cantilever beam, 300 μm long and 44 μm wide, is considered in this case. Therefore, the mechanical parameters for such a cantilever, given the equations previously introduced and the properties reported in Table 2.1, are as follows: cantilever total mass, $M = 9.075 \times 10^{-11}$ Kg; elastic constant, $K = 0.596$ N/m.

The thermal transient analysis of the device requires the determination of the coefficient of convection, h. Furthermore, the assumption of uniform temperature distribution must be verified by means of the Biot number.

Thermal convection is a complex phenomenon involving complicated mathematical models. In practical applications, some empirical correlations between dimensionless parameters are used. There are many correlations, each valid for certain system configurations, geometries, fluid properties and in different ranges of thermo-physical parameters. From this point of view, the cantilever could be considered as a horizontal plate experiencing free convection through the surrounding air. The geometry of the system is taken into account in the characteristic length, L_c. Fluid properties, such as density, viscosity, etc., are temperature-dependent, but for the sake of this analysis, they are evaluated at a reference temperature, called the 'fluid temperature', θ_f, which is defined as the arithmetic mean between the solid surface temperature, θ_s and the undisturbed fluid temperature, θ_∞.

The cantilever is made of silicon oxide and a metal and thus the maximum allowed temperature is the melting temperature of the metal (an aluminium alloy), which is about 730 K; therefore, by assuming θ_0 close to $\theta_\infty (\sim 300 - 350 \,\text{K})$, a maximum $\Delta\theta$ of 350 K will be taken into account here. If $\theta_\infty = 300 \,\text{K}$, it results in $\theta_s = 650 \,\text{K}$ and then $\theta_f = 475 \,\text{K}$. The thermo-physical properties of air are tabulated in Incoprera and De Witt (2001) for several temperature values, by taking values for θ_f and the Rayleigh number results, $Ra = 1.97 \times 10^{-4}$, where the Rayleigh number is defined as:

$$Ra = \frac{g\beta \left(\theta_s - \theta_\infty\right) L_c^3}{\nu\alpha} \tag{2.34}$$

where g is the gravitational field, β the volumetric thermal expansion coefficient, ν the viscosity, α the thermal diffusivity and L_c the characteristic length. Due to the small value of the Rayleigh number, the following correlation can be applied for the *Nusselt number*:

$$Nu = 0.54 Ra^{0.25} \tag{2.35}$$

with the Nusselt number being defined as:

$$Nu = \frac{hL_c}{k_f} \tag{2.36}$$

where k_f is the thermal conductivity of the fluid. The resulting value for h, calculated from equation (2.36), is 67.6 W/m²/K.

The lumped thermal capacitance method can be applied if $Bi \ll 1$. As shown in equation (2.30), in the computation of Bi the thermal conductivity k of the solid material is involved; due to the inhomogeneous structure of the cantilever a sort of 'thermal equivalent material' should be defined. The cantilever is made of a high-thermal-conductivity material, i.e. aluminium, and a very poor conductivity material, i.e. silicon dioxide. The worst condition is obtained if one considers the whole cantilever made of silicon dioxide. In this case, due to the dimensions of the considered devices, the Biot number results in $Bi = 8.034 \times 10^{-5}$; thus, even in the worst case the Biot number is significantly smaller than 1 – then, the 'lumped' thermal-capacitance method can be applied. By assuming averaged thermal properties, the thermal time constant results, from equation (2.28) as $\tau_{th} = 42 \,\text{ms}$.

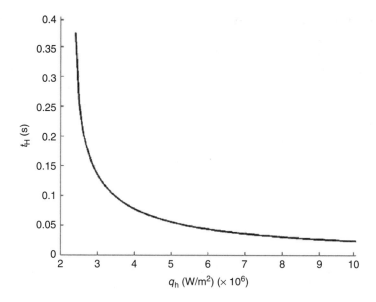

Figure 2.18 Time required by the cantilever to reach a ΔT of 350°C as a function of the incoming power density (Baglio *et al.*, 2002). Reproduced with kind permission from Elsevier

Clearly, the actual time required to reach a given temperature depends also on the power density q_h received over the heated surface area A_h. A plot of the time required by the structure to reach a maximum $\Delta\theta$ of 350°C, t_H (heating time), when the heated area A_h is a circular spot with a diameter equal to the cantilever width (44 μm), is shown in Figure 2.18.

The cooling time t_C can be defined as the time required by the temperature of the structure to decrease from the maximum value to the initial $\Delta\theta = \theta_0 - \theta_\infty \approx 0$; such a time depends only on the maximum temperature reached and can be obtained by solving equation (2.32) when $q_h = 0$ (the source is turned off) and with the initial condition $\Delta\theta(0) = \theta_0 - \theta_\infty = 350°C$; this results in the following:

$$t_C = \tau_{th} \log\left(\frac{\Delta\theta_{max}}{\Delta\theta(0)}\right) \tag{2.37}$$

By substituting the numerical values, equation (2.37) gives: $t_C = 149$ ms.

As expected, the cooling transient dominates the dynamics of the system and therefore thermal actuators are best-suited when large forces or displacements are required in static or quasi-static operation. However, miniaturization helps in improving speed, especially if more efficient heat-dissipation mechanisms and paths are involved (for example, conduction through a solid or forced convection).

For the purpose of completion, the example provided here, if a power density $q_h = 5.5 \times 10^6 1/Wm^2$ is assumed, from Figure 2.18 results in $t_H \approx 50$ ms. Then, the minimum operating period results $T = t_h + t_c \approx 200$ ms and thus the maximum operating frequency is $f_{max} = 5$ Hz. In the considered example, the power required by the actuator to perform the desired work in a given time (50 ms) is $P = q_h A_h = 8.36$ mW.

The scaling laws relative to the proposed thermomechanical actuators will now be briefly analysed. The displacement of a bilayer cantilever tip consequent to a temperature change is expressed by equation (2.29). The temperature variation, $\Delta\theta$, as a function of time is expressed by equation (2.32). The maximum value of temperature variation achieved at the end of the heating phase is given by the following:

$$\Delta\theta_{max} = \frac{q_h A_h}{h A_c} \tag{2.38}$$

The ratio between the heated area, A_h, and the cooling area, A_c, does not scale with linear dimensions. The heating power density, q_h, depends on the source and then it does not need to be considered at this level of analysis. The coefficient of convection, h, depends on the characteristic length, and then on the linear dimensions of the system. Starting from equation (2.36) and substituting backwards gives:

$$h = \frac{k_f}{L_c} Nu = 0.54 \frac{k_f}{L_c} Ra^{0.25} = 0.54 \frac{k_f}{L_c} \left[\frac{g\beta (\theta_s - \theta_\infty) L_c^3}{\nu\alpha} \right]^{0.25} \tag{2.39}$$

Equation (2.39) shows that h is proportional to $L_c^{-1/4}$ – thus it scales as $l^{-1/4}$ – then $\Delta\theta_{max}$ scales as $I^{1/4}$ and δ as $I^{5/4}$.

This means that, if the linear dimensions of the system are proportionally scaled down by a factor of 10, for example, the coefficient of convection, h, increases by a factor 1.778. This result can be explained by the increased surface-to-volume ratio of the scaled system, which favours the convective heat exchange.

The maximum temperature variation, $\Delta\theta_{max}$, scales by a factor of 0.56 – thus it is reduced. This result is obvious because the energy density of the source has not been modified, and then the scaled system collects a smaller amount of energy due to its reduced dimensions (volume or surface, depending on the excitation mechanism involved). Moreover, the increase of h in the scaled system leads to a higher dispersion of heat during the heating phase.

Finally, the tip displacement, δ, scales by a factor of 0.056. Clearly, the absolute displacement in a scaled actuator is smaller than in the non-scaled one, because the absolute linear dimensions are smaller and the materials are supposed to be the same. However, the relative displacement, or the displacement normalized to the cantilever length, scales more favourably than the absolute displacement.

ACKNOWLEDGEMENTS

1. Section 2.2.1 'An interesting classification... the reluctances of gap and core are equal.', pp. 26–31. Portions of the text are reproduced from Z. Nami, C.H. Ahn and M.G. Allen (1996). An energy-based design criterion for magnetic microactuators, *J. Micromech. Microeng.*, 6, 337–344 and are reproduced by permission of IOP Publishing.
2. Section 2.2.2 'Ferromagnetic materials have proven to be... and typical currents are considered.', pp. 31–36. Portions of the text are reproduced from J.W. Judy, H. Yang, N.V. Myung, K.C.-K. Yang, M. Schwartz and K. Nobe (2000). Integrated ferromagnetic microsensors and microactuators, in *Proceedings of the Fifth International Symposium on Magnetic Materials, Processes and Devices*, 198th Meeting of the Electrochemical Society, Phoenix, AZ, USA, October 22–27, pp. 456–68 and are reproduced by kind permission of The Electro-chemical Society.
3. Section 2.3.1 'The analysis of operation and scaling for... favourably than the absolute displacement.', pp. 40–50. Portions of the text are reproduced from S. Baglio, S. Castorina, L. Fortuna and N. Savalli (2002). Modeling and design of novel photo-thermo-mechanical microactuators, *Sensors and Actuators A: Phy.*, 101, September 185–193 and are reproduced by kind permission of Elsevier.

REFERENCES

C.H. Ahn and M.G. Allen (1992). A fully integrated micromagnetic actuator with a multilevel meander magnetic core, in *IEEE Solid-State Sensor and Actuator Workshop (Hilton Head) Technical Digest*, pp. 14–18.

C.H. Ahn, Y.J. Kim, and M.G. Allen (1993). A planar variable reluctance magnetic micromotor with fully integrated stator and wrapped coils, in *Proceedings of IEEE Microelectromechanical Systems (MEMS '93)*, Fort Lauderdale, FL, USA, February 7–10, pp. 1–6.

B. Andò, S. Baglio, A. Bulsara and V. Sacco (2005). RTD based Fluxgate magnetometers, *IEEE Sensor J.*, 5, 895–904.

M. Ataka, A. Omodaka, N. Takeshima and H. Fujita (1993). Fabrication and operation of polyimide bimorph actuators for a ciliary motion system, *J. Microelectromech. Syst.*, 2, 146–150.

S. Baglio, L. Latorre and P. Nouet (1999). *Development of Novel Magnetic Field Monolithic Sensors with Standard CMOS Compatible MEMS Technology*, SPIE'99, Los Angeles, CA, USA.

S. Baglio, S. Castorina, L. Fortuna and N. Savalli (2002). Modeling and design of novel photo-thermo-mechanical microactuators, *Sensors and Actuators A: Phy.*, **101**, 185–193.

J.C. Chiou and Y.J. Lin (2005). A novel large displacement electrostatic actuator: prestress comb-drive actuator, *J. Micromech. Microeng.*, **15**, 1641–1648.

T. Ebefors, J.U. Mattson, E. Kalvesten and G. Stemme (1999). A robust micro conveyor realized by arrayed polyimide joint actuators, in *Proceedings of IEEE International Conference on Micro Electro Mechanical Systems (MEMS'99)*, Orlando, FL, USA, January 1999, pp. 17–21.

M. Gad-el-Hak (Ed.) (2001). *The MEMS Hand book*, CRC Press, Boca Raton, FL, USA.

J.M. Gere and S.P. Timoshenko (1997). *Mechanics of Materials*, 4th Edition, PWS, Boston, MA, USA, Ch. 9, pp. 599–680.

Y.B. Gianchandani and K. Najafi (1996). Bent beam strain sensors, *J. Microelectromech. Syst.*, **5**, 52–58.

H. Guckel, T.R. Christenson, H.J. Skrobis, T.S. Jung, J. Klein, K.V. Hartojo and I. Widjaja (1993). A first functional current excited planar rotational magnetic micromotor, in *Proceedings of IEEE Microelectromechanical Systems (MEMS)*, pp. 7–11.

T. Hirano, T. Furuhata, K.J. Gabriel, and H. Fujita (1992). Design, fabrication, and operation of submicron gap comb-drive microactuators, *IEEE J. Microelectromech. Syst.*, **1**, 52–59.

R. Holzer, I. Shimoyama and H. Miura (1995). Hybrid electrostatic-magnetic microactuators, *Proc. 1995 IEEE Int. Conf. Robot Automat. (R&A '95)*, Vol. 3, Nagoya, May 21–27, pp. 2941–2946.

F.P. Incropera, D.P. De Witt (2001). *Fundamentals of Heat and Mass Transfer*, 5th Edition, John Wiley and Sons, Inc., New York, NY, USA.

J.W. Judy, H. Yang, N.V. Myung, K.C.-K. Yang, M. Schwartz and K. Nobe (2000). Integrated ferromagnetic microsensors and microactuators, in *Proceedings of the Fifth International Symposium on Magnetic Materials, Processes and Devices*, 198th Meeting of the Electrochemical Society, Phoenix, AZ, USA, October 22–27, pp. 456–68.

L. Latorre, Y. Bertrand and P. Nouet (1998). On the use of test Structures for the electromechanical characterization of a CMOS compatible MEMS technology, in *Proceedings of IEEE 1998 International Conference on Microelectronic Test Structures*, **11**, pp. 177–182, Kanazawa, Japan, March, pp. 23–26.

Z. Nami, C.H. Ahn and M.G. Allen (1996). An energy-based design criterion for magnetic microactuators, *J. Micromech. Microeng.*, **6**, 337–344.

S.J. O'Shea, P. Lu, F. Shen, P. Neuzil and Q.X. Zhang (2005). Out-of-plane electrostatic actuation of microcantilevers, *Nanotechnology*, **16**, 602–608.

M. Okyar, X.-Q. Sun and W.N. Carr (1997). Thermally excited inchworm actuators and stepwise micromotors: analysis and fabrication, *Proc. SPIE*, **3224**, 372–379.

F. Paschen (1889). *Ann. Physik*, **37**, 69–96.

L. Que, J.-S. Park and Y.B. Gianchandani (2001). Bent beam electrothermal actuators – Part I: single beam and cascaded devices, *J. Microelectromech. Syst.*, **10**, 247–254.

W. Riethmuller and W. Benecke (1988). Thermally excited silicon microactuators, *IEEE Trans. Electr. Dev.*, **35**, 758–763.

P. Ripka, S. Kawahito, S.O. Choi, A. Tipek and M. Ishida (2001). Micro-Fluxgate sensor with closed core, *Sensors Actuators A*, **91**, 65–69.

M.A. Rosa, S. Dimitrijev and H.B. Harrison (1998). Enhanced electrostatic force generation capability of angled comb finger design used in electrostatic comb-drive actuators, *Electron. Lett.*, **34**, 1787–1788.

J.W. Suh, R.B. Darling, K.-F. Bohringer, B.R. Donald, H. Baltes and G.T.A. Kovacs (1999). CMOS integrated ciliary actuator array as a general-purpose micromanipulation tool for small objects, *J. Microelectromech. Syst.*, **8**, 483–496.

X.-Q. Sun, X. Gu and W. N. Carr (1996). Lateral in-plane displacement microactuators with combined thermal and electrostatic drive, in *Proceedings of Solid-State Sensor and Actuator Workshop*, Hilton Head, SC, USA, June, pp. 152–155.

W.C. Tang, T.H. Nguyen and R.T. Howe (1989). Laterally driven poly-silicon resonant microstructures, in *IEEE Microelectromechanical Systems Workshop Technical Digest*, Salt Lake City, UT, USA, February 20–22, pp. 53–59.

A.R. von Hippel (1959). *Molecular Science and Molecular Engineering*, The Technology Press of M.I.T and John Wiley & Sons, Inc., New York, NY, USA, pp. 39–47.

B. Wagner and W. Benecke (1991). Microfabricated actuator with moving permanent magnet, *Proc. MEMS 1991*, Nara, pp. 27–32.

J.G. Webster (Ed.) (1998). *The Measurement, Instrumentation and Sensors Handbook* (Electrical Engineering Handbook Series), CRC Press, Boca Raton, FL, USA.

D.C. White and H.H. Woodson (1959). *Electromechanical Energy Conversion*, John Wiley & Sons, Inc., New York, NY, USA.

K. Yanagisawa, A. Tao, T. Ohkobu and H. Kuwano (1991). Magnetic microactuator, *Proc. MEMS 1991*, Nara, pp. 120–124.

3

Scaling of Thermal Sensors

3.1 THERMOELECTRIC SENSORS

As a 'concrete' example of thermal sensors, the analysis, design and scaling of integrated thermocouples for the measurement of temperature will be addressed here. Thanks to their intrinsically simple structure and effective principles of operation, thermoelectric sensors can potentially be implemented in any IC technology in which at least two different conductive materials are available. Thermocouples or thermopiles can then be designed with minimal added efforts and area consumption. The effectiveness of the approach depends on the thermoelectric properties of the materials used. However, the realization of actual thermoelectric devices requires the careful optimization of some design parameters in order to achieve high performance (sensitivity, efficiency) with minimum area consumption.

The *Seebeck effect* in semiconductors and its applications to integrated thermocouples and thermopiles have been largely addressed in the literature (Van Herwaarden, 1984; Van Herwaarden and Sarro, 1986; Akina *et al.*, 1998). However, a brief review of the thermoelectric effect's theory will be given in the following.

The Seebeck or *Thermoelectric* effect is the generation of an electrical voltage due to a temperature difference between a 'weld' of two different materials, say a thermocouple, and the other ends of these wires, as shown in Figure 3.1. The generated thermoelectric voltage, ΔV, is related to the temperature difference, ΔT, by a linear relationship, as a first

Scaling Issues and Design of MEMS   S. Baglio, S. Castorina and N. Savalli
© 2007 John Wiley & Sons, Ltd

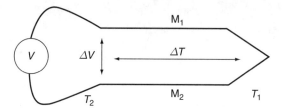

Figure 3.1 Illustration of the Seebeck effect (Baglio *et al.*, 2003). Reproduced with kind permission of TIMA Laboratory

approximation, by mean of the relative Seebeck coefficient or relative thermopower α_{12}:

$$\Delta V = \alpha_{12}\Delta T \qquad (3.1)$$

The linear dependence expressed by equation (3.1) is often enough accurate; however, sometimes a quadratic approximation could be required.

If N thermocouples are connected in series and are all subjected to the same temperature difference, ΔT, the total thermoelectric voltage, ΔV, is the sum of the voltages generated by each thermocouple, given by equation (3.1). If all of the thermocouples are the same, this results in the following:

$$\Delta V = N\alpha_{12}\Delta T \qquad (3.2)$$

The series connection of N thermocouples is called a *thermopile*.

The relative Seebeck coefficient of the two materials used to form the thermoelectric element must be as high as possible, in order to obtain the highest measurement sensitivity. As a metric to evaluate the performance of thermocouples, a figure of merit Z has been defined as follows:

$$Z = \frac{\alpha_{12}^{2}}{\kappa\rho} \qquad (3.3)$$

where κ is the total thermal conductivity and ρ is the electrical resistivity. The so-defined figure of merit Z can be used to optimize the design.

The measurement reliability of a thermocouple can be affected by thermal noise – the Peltier effect that arises if an excessive electrical current flows through the thermocouple, i.e. if the measuring instrument shown in Figure 3.1 has an input impedance which is not high enough.

In standard CMOS technology, the materials available for the realization of thermocouples are p- and n-doped Si, available as diffused wells in the substrate, polycrystalline Si and metallization layers. It has been shown than Si, and semiconductors in general, show the highest values of the Seebeck coefficient and that this depends on the doping levels (Geballe and Hull, 1955).

Some device prototypes have been realized and characterized, and the results will be reported in this chapter. The devices proposed here are thermopiles realized with the 0.8 μm CMOS CXQ technology by 'Austria Mikro Systeme' (AMS). Two thermopiles have been designed by using the metal 1/p$^+$ diffusion (made of 14 thermocouples) and metal 1/poly 1 layers (made of 18 thermocouples). Thermopiles run along one side of a 3.1×3.1 mm^2 die. At both thermocouple ends, heating resistors have been realized to allow the characterization of the devices.

The complete die layout is reported in Figure 3.2, while in Figure 3.3 details of one thermopile joint are reported.

Simulations of the devices have been performed with 'ANSYS$^®$'. Details of the temperature distribution due to the current flowing in the heating resistor are shown in Figure 3.4.

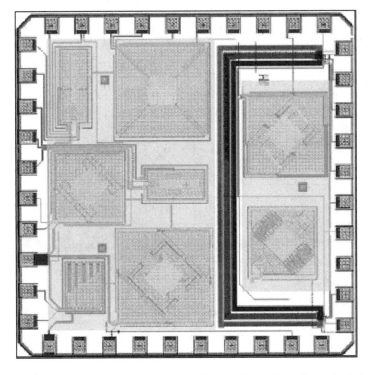

Figure 3.2 Layout of the die. The thermopiles are the 'C-shaped', unshaded structures on the right-hand side (Baglio *et al.*, 2003). Reproduced with kind permission of TIMA Laboratory

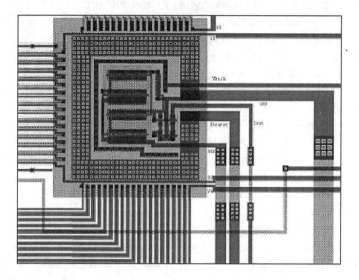

Figure 3.3 Details of one of the thermopile joint layouts. The thermopiles are made of 14 metal 1/p$^+$ diffusion thermocouples (left-hand side) and 18 metal 1/poly 1 thermocouples (bottom), respectively. A heating polysilicon resistor is embedded to allow device characterization (Baglio *et al.*, 2003). Reproduced with kind permission of TIMA Laboratory

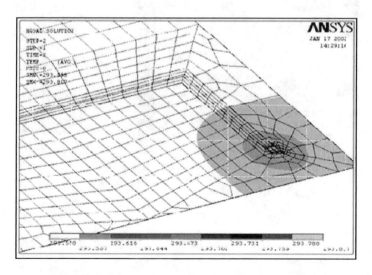

Figure 3.4 Simulated temperature distribution around the thermopile's hot joint (Baglio *et al.*, 2003). Reproduced with kind permission of TIMA Laboratory

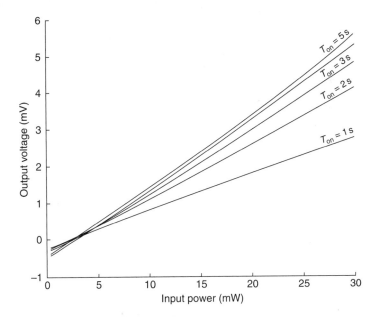

Figure 3.5 Metal 1/poly 1 thermopile (18 thermocouples) response to heating power for different values of the heating time (Baglio *et al.*, 2003). Reproduced with kind permission of TIMA Laboratory

In order to verify the operation of integrated thermocouples, some prototypes have been designed and realized by using a standard CMOS technology, as shown in Figure 3.2, and they have been characterized in terms of the power dissipated on the integrated heating resistor, for different values of the operating time (in other words, the characterization has been made in terms of energy). The results are shown in Figures 3.5 and 3.6 for the metal 1/poly 1 and the metal 1/p$^+$ thermopiles, respectively.

A very good repeatability has been observed among several thermopiles samples, on different dies and under different measurement conditions, confirming the robustness of this approach to process parameters and environment conditions variations.

In order to obtain the response of the thermopiles in terms of the 'actual' on chip temperature, the values of the Seebeck coefficients are needed. As a first approximation, the Seebeck coefficient for metal/'poly' thermocouples realized with the same CMOS technology by Baltes and coworkers has been taken into account (Baltes *et al.*, 1998). The value reported for this coefficient is $\alpha_{\text{met/poly}} = 109 \, \mu\text{V/K}$. Therefore, the actual temperature difference between the thermopile joints can be estimated by solving equation (3.2) for the temperature difference and by substituting

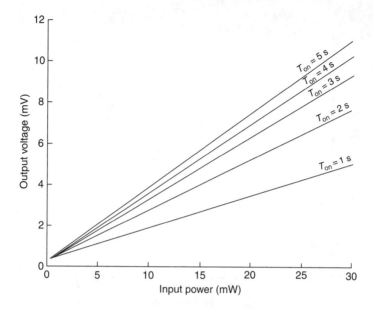

Figure 3.6 Metal $1/p^+$ diffusion thermopile (14 thermocouples) response to heating power for different values of the heating time (Baglio *et al.*, 2003). Reproduced with kind permission of TIMA Laboratory

the reported Seebeck coefficient and the measured output voltage. The results of these estimations are reported in Figure 3.7.

By considering, for example, the plot for $T_{on} = 4\,s$ in Figure 3.7, which is the one showing the strongest linear behaviour, this is fitted by the following relationship between the input power P and the temperature difference ΔT:

$$\Delta T = 0.1P - 0.17 \qquad (3.4)$$

Since the thermopiles are thermally connected in parallel, and the heaters are the same, it follows that for a given value of the input power, both of the thermopiles are subjected to the same temperature difference. Therefore, equation (3.4) is valid for both types of thermopiles and can be used to estimate the response in terms of temperature difference of the metal $1/p^+$ diffusion thermopiles. The resulting average Seebeck coefficient for this type of thermocouple is $\alpha_{met/p^+} = 257.83\,\mu V/K$.

By using the estimated Seebeck coefficients and the measured data, it is now possible to plot the characteristics of the two types of thermopiles in terms of output voltage as a function of the temperature difference between the joints. The results are reported in Figures 3.8 and 3.9 for the metal 1/poly 1 and metal $1/p^+$ diffusion thermopiles, respectively.

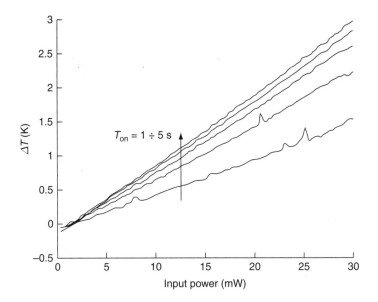

Figure 3.7 Estimated temperature – power relations for the metal 1/poly 1 thermopile (18 thermocouples) (Baglio *et al.*, 2003). Reproduced with kind permission of TIMA Laboratory

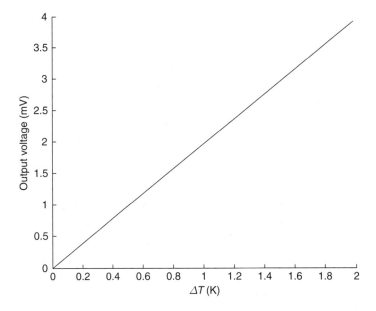

Figure 3.8 Voltage-to-temperature characteristics of the metal 1/poly 1 thermopile (18 thermocouples) (Baglio *et al.*, 2003). Reproduced with kind permission of TIMA Laboratory

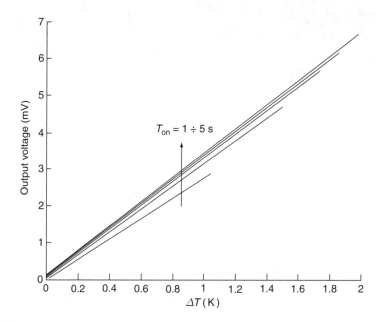

Figure 3.9 Voltage-to-temperature characteristics of the metal $1/p^+$ diffusion thermopile (14 thermocouples) (Baglio *et al.*, 2003). Reproduced with kind permission of TIMA Laboratory

The plot in Figure 3.8 has been obtained by superimposing the single plots for the different values of T_{on} – its slope corresponds to a Seebeck coefficient value, for the single thermocouple, of about $107\,\mu V/K$, which is, in practice, the value taken from the literature. On the other hand, data for the metal $1/p^+$ diffusion thermocouple, in Figure 3.9, show variations in the slope of the characteristics, resulting in variations of the Seebeck coefficient with the heating time and thus with the maximum temperature reached. This behaviour could be explained as a higher sensitivity of the Seebeck coefficient to temperature value, for semiconductors. Qualitatively, it is known that in semiconductors the Seebeck coefficient is a function of the doping level and therefore, since the carrier concentration is also a function of the temperature, it follows that the Seebeck coefficient is also a function of temperature. However, the estimated value of the Seebeck coefficient for the single metal $1/p^+$ diffusion thermocouple, from the data relative to $T_{on} = 5\,s$, results in $256.4\,\mu V/K$, and is thus very close to the average value estimated by using equation (3.4).

These examples have shown the feasibility and effectiveness of integrated thermal sensors exploiting the thermoelectric Seebeck effect. Thermopiles and thermocouples can be scaled and integrated in almost any IC technology, with a very slight increase in design complexity and area consumption. Substrate micromachining processes may result in enhanced integrated device performance, thanks to the improved thermal insulation; however, these additional steps may not be necessary for the device operation, as the reported examples have shown. Therefore, thermoelectric sensors represent a feasible alternative for the miniaturization of temperature, and more generally, thermal sensors.

3.2 APPLICATION: DEW-POINT RELATIVE HUMIDITY SENSORS

In this section, another example of a device exploiting thermoelectric effects to perform sensing on the microscale is presented. The device presented here is an integrated dew-point relative humidity sensor realized by adopting standard CMOS technology for applications in various fields.

Both the Peltier and the Seebeck thermoelectric effects are exploited in the operation of the device: micro-Peltier cells are used to cool down a plate which is suspended over the substrate by means of bulk micromachining to improve thermal insulation and reduce the thermal mass to be cooled, thus exploiting the advantages of scaling thermal phenomena. Integrated thermocouples are used to measure the temperature difference between the plate and the environment at the dew-point condition. The dew-point temperature is that at which moisture appears at the functional surface, i.e. the plate. The cooling process has to be stopped when the dew point is reached; therefore, a sensing technique is needed to detect the presence of water over the plate. An interdigited capacitive sensor is used for this purpose. The presence of moisture results in a change of permittivity between the fingers which, in turn, results in a capacitance variation. The capacitive sensor output is used to stop the cooling process and trigger the temperature measurement. The relative humidity at atmospheric pressure is a function of the dew point and environmental temperature values.

The analysis and design of a device prototype are presented in the following, together with some simulated and experimental results.

The choice falls on a relative humidity sensor because the measurement and control of this parameter is of strategic importance in many application fields, like in environmental pollution monitoring, in air conditioning, meteorology and industry. For example (Moghavemmi et al., 2000), in the food industry it is very important to maintain both relative humidity and temperature within well-defined ranges, in order to keep the products in their best conditions; also in the pharmaceutical industry the control of relative humidity is critical in the production process. In such application fields, and in many others, miniaturized, low-cost, fast, high-sensitivity relative humidity sensors are needed (Pascal-Delannoy *et al.*, 1998).

Some of the already existing approaches for the measurement of relative humidity include piezoresistive, resistive (Buchold *et al.*, 1998) and capacitive sensors and quartz oscillators (Pascal-Delannoy *et al.*, 2000]. However, approaches based on dew-point temperature detection(Pascal-Delannoy *et al.*, 1998) are widely used, thanks to their high intrinsic reproducibility. The operation of dew-point sensors is typically based on the cooling of a given surface until a water film occurs; then, the dew point and ambient temperatures are measured and used to extract information on the relative humidity. Different approaches are characterized by the method used to detect the moisture, for example, surface acoustic waves (Galipeau *et al.*, 1995; Hoummady *et al.*, 1995), resonant quartz oscillators (Pascal-Delannoy *et al.*, 2000), chilled mirrors (Pascal-Delannoy *et al.*, 1998; Sorly *et al.*, 2002], whereas Thermo-Electric Coolers (TECs) are typically used to cool the sensing surface.

Commercially available relative humidity sensors have typical dimensions of some cubic centimetres or greater and are characterized by an operating range from 20% up to 95%, response times of the order of some seconds and a current consumption as high as 2.5 A (Pascal-Delannoy *et al.*, 2000). Miniaturization of relative humidity sensors may overcome these problems. Moreover, the realization of such sensors in standard CMOS technology allows large-volume and low-cost fabrication of microsensors and electronics on the same chip. Integration also leads to faster sensors, due to the smaller dimensions, and higher noise and interference immunity, due to the reduced number of 'off-chip' connections. However, from a strategic point of view, miniaturization has the potential advantage of allowing novel applications, especially in those cases where there exist strong space constraints, or where a large number of sensors is required.

3.2.1   Device structures and operating principles

The operating principles of the proposed sensor are those of dew-point relative humidity sensors, with a capacitive detection of the condensed-water film. The device is composed of a suspended plate which is cooled by means of integrated TECs. The plate is suspended by four arms to improve thermal insulation. These arms are used to mechanically sustain the sensor plate, to provide a mechanical support for the TECs and, at the same time, they must guarantee a high enough thermal resistance. Integrated thermocouples are also used to measure the temperature of the plate at the dew point plus the ambient temperature. Interdigited capacitors connected in a bridge configuration are realized on top of the sensing plate to allow water-film detection. The presence of such a water film over the capacitors will unbalance the bridge, whose output signal can be used as a trigger for the temperature reading. Since thermocouples provide only an output voltage proportional to a temperature difference, an absolute temperature reference has also been integrated in to the sensor design.

A schematic of the proposed device, together with the operating principles of the measurement system, is shown in Figure 3.10.

Although the structure of the proposed device could be generalized, the modelling and the successive design choices are strictly related to the technology adopted for the device realization. In fact, the choice of a given technology determines the number and type of available materials, together with their thermo-physical properties, and the degrees of freedom the designer has in conceiving the final device structure.

The technology chosen for this work is the $0.8\,\mu m$ CMOS CXQ from AMS. In such a technology, the silicon substrate and the oxide layers are available as structural materials, to realize the plate and the four arms, while one polysilicon and two metal layers are the functional materials, which have been used to realize the TECs, the thermocouples and the interdigited capacitor. Thermal isolation of the plate from the substrate can be improved with a selective etching of the substrate by means of anisotropic etching of silicon in TMAH solutions.

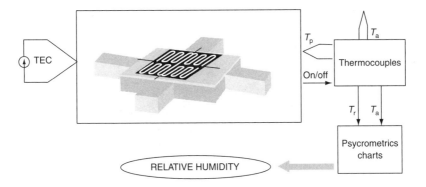

Figure 3.10 Schematic, along with the operating principles, of the proposed system (Savalli *et al.*, 2004). Reproduced with kind permission of SPIE

3.2.2   Device modelling and simulations

The relative humidity sensor proposed here can be viewed as composed
of three basic functional blocks: the cooling system, which includes both
the TECs, and the mechanical structure of the device (the plate and the
four arms) which represents the thermal load for the TECs, while the
other two functional blocks are the thermocouples and the interdigited
capacitive sensor. The absolute temperature reference used in this work
has been addressed in Baglio *et al.* (2003).

The cooling system represents the main issue in the proposed device. It
exploits the Peltier effect to subtract heat from the plate, thus leading to a
local temperature decrease. This effect is partially counterbalanced from
heating fluxes due to the Joule effect in the TECs, thermal conduction
from the substrate through the four arms and thermal convection from
the surrounding air. The heat fluxes involved in the operation of the
sensor's cooling system are represented in Figure 3.11.

The energy balance at the plate of Figure 3.11 yields:

$$U' = Q_P - Q_J - Q_{cd} - Q_{cv} \tag{3.5}$$

where U' is the rate of variation of the internal energy of the system, Q_P
is the outgoing heat flux due to the Peltier effect and Q_J, Q_{cd} and Q_{cv} are
the incoming heat fluxes due to, respectively, the Joule effect, thermal
conduction and thermal convection. By developing equation (3.5) (Sorly
et al., 2002), gives:

$$-\rho c V \frac{dT_c}{dt} = N\sigma_{ab}IT_c - R_{el}I^2 - Nk_{th}(T_h - T_c) - hA_c(T_\infty - T_c) \tag{3.6}$$

where ρ is the density of the plate, c is its specific heat and V its volume,
T_c is the temperature at the cold junction, N is the number of Peltier cells,

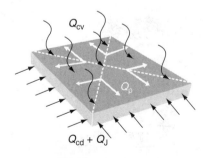

Figure 3.11   Heat fluxes involved in the sensor operation (Savalli *et al.*, 2004).
Reproduced with kind permission of SPIE

σ_{ab} is the Seebeck coefficient, I is the current intensity, R_{el} is the electrical resistance of the TECs, k_{th} is the thermal conductivity of the arms, T_h is the temperature at the hot junction, h is the convective coefficient, A_c is the plate area and T_∞ is the temperature of the unperturbed air (which is supposed to be equal to the ambient temperature).

Equation (3.6) has been used to implant a MATLAB® script to simulate the behaviour of the sensor's cooling system. This takes into account the main thermal effects involved in the operation of the device. However, it has also been used to generate an equivalent circuit model of the cooling system to be simulated in SPICE (Chavez et al., 2000), which allows taking into account several 'parasitic' effects in a easier way. The equivalent circuit model of the device is shown in Figure 3.12.

In the equivalent circuit of this figure, the hot and cold junctions of the TECs have been represented. In such a circuital model, the temperatures and heat fluxes have been represented as voltages and currents, respectively. The Peltier effect and thermal convection contributions at each junction are modelled through voltage-controlled current sources, while the two independent current sources model the Joule effect contribution and the resistors and capacitors represents the thermal resistances and capacitances, respectively.

The cooling system is critical to the device performance as it has to remove the larger amount of heat from the suspended plate with minimal losses. In the process of cooling the suspended plate, losses are represented by the heat fluxes entering the plate, for example, the Joule heat dissipation on the TECs, thermal conduction from the higher-temperature substrate and thermal convection from the surrounding air. All of these contributions are taken into account in the right-hand side of equation (3.6). The heat flux removed from the plate, thanks

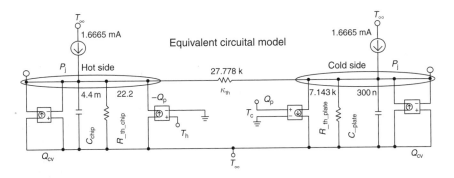

Figure 3.12 Equivalent electrical circuit of the sensor's cooling system (Savalli et al., 2004). Reproduced with kind permission of SPIE

to the Peltier effect, is proportional to the number of TECs, N, and to the current, I, flowing through them. However, these parameters affect also the Joule dissipation and the other thermal loss mechanisms as well. In fact, the number and shape of the TECs directly influence their electrical and thermal resistances, which are key parameters in the loss mechanisms. The design of a high-performance device thus requires a careful optimization among these parameters.

The current is limited by the Joule effect which produces heating. The number of thermo-electric elements which can be used is mainly limited from geometric constraints. In fact, a larger number of elements would require a larger plate, and thus a higher thermal load, and larger arms, which would mean higher losses due to thermal conduction (i.e. a lower thermal resistance). The widths of the thermo-electric elements cannot be reduced under a certain value to reduce area consumption because then the Joule heating may become dramatic. The problem of the higher electrical resistance of a larger number of thermo-electric elements can be overcome by connecting them electrically in parallel, instead of in series.

The main loss contribution due to thermal conduction is expected to be due to the four suspending arms, which thermally connect the plate to the silicon substrate, and to the thermo-electric elements themselves, because they are made of thermally conductive materials. Therefore, the geometry of such arms and thermo-electric elements is an issue in the design of the device.

To find the optimal solution among the above mentioned 'tradeoffs', the minimum temperature reached at the cold junction in the steady state has been simulated as a function of the lengths of the thermo-electric elements for different values of N. In such a simulation, the width of the metal 1 side of the thermo-electric elements has been chosen to be equal to the minimum allowed by the technology; as a consequence, the maximum current and the width of the poly 1 side have been determined. The results of the simulation are reported in the plot of Figure 3.13.

In the plot of Figure 3.13, it is shown that the minimum temperature at the cold junction decreases for a higher number of TECs. It also shows that for a given N, there exists an optimal length value for the TECs, which corresponds to the minimum temperature. This optimal length corresponds to that value under which there is not enough thermal insulation from the substrate, and above which thermal losses become dominant.

The capacitive detection of condensed water over the plate is proposed in this approach to trigger the measurement of the dew-point temperature. The capacitive sensor is realized with an interdigited structure on top of the sensor plate, by exploiting the metal 2 layer. The

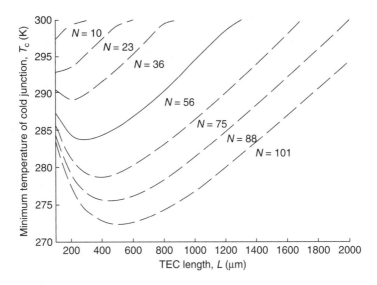

Figure 3.13 Simulated minimum temperature at the cold junction, T_c, versus the length of the thermo-electric coolers (TECs), L, and number, N (Savalli *et al.*, 2004). Reproduced with kind permission of SPIE

proposed strategy is to realize a capacitive, four-branch bridge, with two branches covered with the pad layer, passivation dielectric used in the chosen technology, and the other two branches with metal 2 exposed to air. In such a way, the exposed branches experience a change in the dielectric medium, from air to water, when the dewpoint is reached, thus resulting in a sudden change of the bridge output signal, which can be used to trigger the temperature measurement.

Both the dew-point and ambient temperatures are read by means of integrated thermocouples, realized with the same materials used for the TECs, poly 1 and metal 1, but with less geometric constraints since they do not have to carry high currents.

3.2.3 Device design

The operating range of a dew-point relative humidity sensor depends on the temperature decreasing the cooling system can achieve with respect to the ambient temperature, at a given pressure, e.g. at atmospheric pressure. Once the dew-point and ambient temperatures are known, the relative humidity value can be achieved from a psycrometric chart or from suitable tables, like the example reported in Table 3.1

Table 3.1 Dew-point temperature as a function of the relative humidity and the ambient temperature (Savalli *et al.*, 2004). Reproduced with kind permission of SPIE

$T_a(K)$	Relative humidity (%)							
	30	40	50	60	70	80	90	95
303.0	283.5	287.9	291.4	294.4	296.9	299.2	301.2	302.1
302.0	282.7	287.0	290.5	293.4	296.0	298.2	300.2	301.1
301.0	281.8	286.1	289.6	292.5	295.0	297.2	299.2	300.1
300.0	281.0	285.2	288.7	291.6	294.1	296.3	298.2	299.1
299.0	280.1	284.4	287.8	290.6	293.1	295.3	297.2	298.1
298.0	279.2	283.5	286.9	289.7	292.1	294.3	296.2	297.1

(Steinbrenken), where each entry reports the dew-point temperature (in K) at a given relative humidity and ambient temperature. For example, if the maximum achievable temperature difference is $\Delta T = -15\,K$, the sensor can measure relative humidity values in the range from 40 % up to 95 %, for any ambient temperature value in Table 3.1.

A device prototype has been designed in order to achieve the above mentioned performance, thus to measure relative humidity values greater than 40%.

The designed prototypes have a $200 \times 200\,\mu m^2$ wide plate, $500\,\mu m$ long suspending arms, which host a number of 56 TEC elements electrically connected in parallel, and 10 thermocouples; the TECs are supplied by a 56 mA electrical current and their power consumption amounts to 196 mW. The intedigited capacitor has been designed in the metal 2 layer and two design solutions have been adopted: one in which two branches of the capacitive bridge are passivated and the other one with all of the entire bridge passivated.

The layout of the device prototypes has been designed by using the 'MEMSPro v.2.0' software package and details of the device's layout are reported in Figure 3.14, where the entire structure of the sensor is visible, together with the absolute temperature reference (on the right-hand side).

The simulation of the prototype device is shown in Figure 3.13 by the solid line (N = 56). It shows how a length of $500\,\mu m$ would allow cooling the plate down to 285 K with respect to an ambient temperature of 300 K. This simulation has also been verified with the equivalent circuit model of Figure 3.12. Moreover, the equivalent circuit model has been used to estimate the transient behavior of the prototype. This simulation results are reported in Figure 3.15, which shows that the required cooling of $-15\,K$ can be achieved in less than 10 ms.

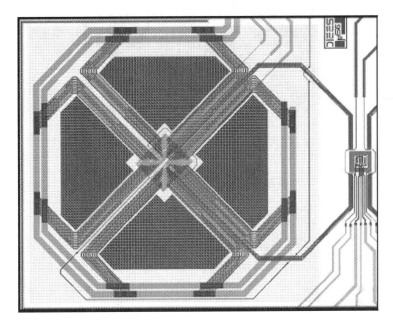

Figure 3.14 Layout of the CMOS integrated dew-point relative humidity sensor. The absolute temperature reference is also visible on the right-hand side (Savalli *et al.*, 2004). Reproduced with kind permission of SPIE

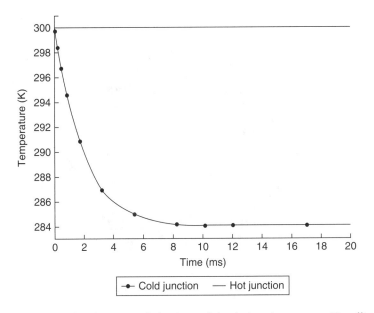

Figure 3.15 Simulated transient behaviour of the designed prototype (Savalli *et al.*, 2004). Reproduced with kind permission of SPIE

Figure 3.16 Microscope image of a device prototype after 'releasing operations' (Savalli *et al.*, 2004). Reproduced with kind permission of SPIE

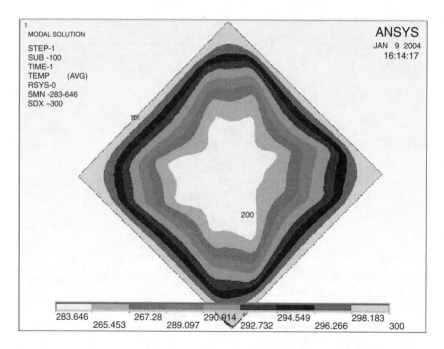

Figure 3.17 'Finite Element Analysis' of the temperature distribution over the plate surface (Savalli *et al.*, 2004). Reproduced with kind permission of SPIE

An image of a final prototype, in which the sensor structure has been suspended by means of anisotropic wet silicon etching in TMAH, is reported in Figure 3.16. As can be seen, the sensor structure has been safely and completely released.

As a further validation of the realized design, simulations with the 'Finite Elements Method' have been performed on the designed prototype. In particular, in Figure 3.17 the regime temperature distribution over the plate surface, when an outgoing heat flux equal to that due to Peltier cooling, together with conductive, convective and Joule heat losses, are taken into account. This simulation also confirms that the temperature in the central area of the plate decreases by $15\,\mathrm{K}$ with respect to the ambient temperature.

3.3 CONCLUSIONS

Simulations have shown that the designed prototypes can measure relative humidity values in the range from 40% up to 95%, at atmospheric pressure and ambient temperatures ranging from $283\,\mathrm{K}$ to $303\,\mathrm{K}$ at least, with response times smaller than $10\,\mathrm{ms}$ and a power consumption of $196\,\mathrm{mW}$. Moreover, finite element analysis confirmed the performances estimated with analytical and circuital models.

The proposed relative humidity microsensor may find potential applications in those fields where a 'reduced-space occupation' is a stringent requirement, or where a distributed, capillary monitoring of the environmental conditions is necessary.

A standard CMOS technology, with an additional bulk micromachining step, has been adopted to design and realize the proposed sensor. This choice allows low production costs and integration with other sensors and electronics.

ACKNOWLEDGEMENTS

1. Section 3.1 'As a "concrete" example of thermal sensors, the . . . . and more generally, thermal sensors.', pp. 53–61. Portions of the text are reproduced from S. Baglio, S. Castorina and N. Savalli (2003). On-Chip Temperature Distribution Monitoring via CMOS Thermocouples, in *THERMINIC 2003, 9th International Workshop on Thermal Investigations of ICs and Systems*, Aix-en-Provence, France, September 24–26, pp. 329–333 and are reproduced by kind permission of TIMA Laboratory.
2. Section 3.2 'The operating principles of the proposed sensor . . . with respect to the ambient temperature.', pp. 62–71. Portions of the text are reproduced from N. Savalli, S. Baglio, S. Castorina, V. Sacco, C. Tringali (2004). Integrated CMOS dew point

sensors for relative humidity measurement, *SPIE International Symposium, Smart Structures and Materials*, Vol. 5389, San Diego, CA, USA, March 14–18, pp. 422–430 and are reproduced by kind permission of SPIE.

REFERENCES

T. Akina, O. Akara, Z. Olgunb and H. Kulahb (1998). An integrated thermopile structure with high responsivity using any standard CMOS process, *Sensors Actuators A: Phys.*, **66**, 218–224.

S. Baglio, S. Castorina and N. Savalli (2003). On-Chip Temperature Distribution Monitoring via CMOS Thermocouples, in *THERMINIC 2003, 9th International Workshop on Thermal Investigations of ICs and Systems*, Aix-en-Provence, France, 24–26 September, pp. 329–333.

H. Baltes, O. Paul and O. Brand (1998). Micromachined thermally based CMOS microsensors, *Proc. IEEE*, **86**, 1660–1668.

R. Buchhold, A. Nakladal, U. Buttner and G. Gerlach (1998). The metrological bahaviour of bimorphic piezoresistive humidity sensors, *Measure Sci. Technol.*, **9** 354–359.

J.A. Chavez, J.A. Ortega, J. Salazar, A. Turò and M.J. Garcia (2000). SPICE Model of Thermoelectric Elements Including Thermal Effects, *in Proceedings of IEEE Instrumentation and Measurement Technology Conference*, Baltimore, MD, USA, 1–4 May, pp. 1019–1023.

D.W. Galipeau, J.D. Stroschine, K.A. Snow, K.A. Vetelio, K.R. Hines and P.R. Story (1995). A study of condensation and dew point using a SAW sensor, *Sensors Actuators B: Chem.*, **24–25**, 696–700.

T.H. Geballe and D.W. Hull (1955). Seebeck Effeect in Silicon, *Phys. Rev.*, **98**, 940–947.

M. Hoummady, C. Bonjour, J. Collin, F. Lardet-Vieudrin and G. Martin (1995). Surface acoustic wave (SAW) dew point sensor: application to dew point hygrometry, *Sensors Actuators B: Chem.*, **26–27**, 315–317.

M. Moghavvemi, K.E. Ng and C.Y. Soo (2000). Remote sensing of relative humidity, in *Proceedings of TENCON 2000*, Kuala Lumpur, Malaysia, 24–27 September, pp. 96–101.

F. Pascal-Dellanoy, A. Sackda, A. Giani, A. Foucaran and A. Boyer (1998). Fast humidity sensor using optoelectronic detection on pulsed Peltier device, *Sensors Actuators A: Phys.*, **65**, 165–170.

F. Pascal-Delannoy, B. Sorli and A. Boyer (2000). Quartz Cristal Microbalance (QCM) used as humidity sensor, *Sensors Actuators A: Phys.*, **84**, 285–291.

N. Savalli, S. Baglio, S. Castorina, V. Sacco, C. Tringali (2004). Integrated CMOS dew point sensors for relative humidity measurement, *SPIE International Symposium, Smart Structures and Materials*, Vol. 5389, San Diego, CA, USA, March 14–18, pp. 422–430.

B. Sorly, F. Pascal-Delannoy, A. Giani, A. Foucaran and A. Boyer (2002). Fast humidity sensors for high range 80–95% RH, *Sensors Actuators A: Phys.*, **100**, 24–31.

T. Steinbrenken. The heatsink guide: Peltier coolers, Website [www.heatsink-guide.com].

A.W. Van Herwaarden (1984). The Seebeck Effect in Silicon ICs, *Sensors Actuators*, **6**, 245–254.

A.W. Van Herwaarden and P.M. Sarro (1986). Thermal Sensors Based on the Seebeck Effect, *Sensors Actuators*, **10**, 321–346.

4

Inductive Sensors for Magnetic Fields

Magnetic field sensors are widely adopted as measuring systems for both direct and indirect measurements (Roumenin, 1994; Kadar *et al.*, 1994; Gardner, 1994). Typical applications consist, for example, in the estimation of the geomagnetic field and biomagnetic field, while a large set of nonmagnetic quantities are often estimated through a preliminary conversion to a magnetic field which is then reconverted in a suitable electric signal by using magnetic field sensors, as in the case of proximity or displacement sensors (Roumenin, 1994).

Regarding magnetic sensors based on the Lorentz law, mentioned in Chapter 2, the auxiliary input which interacts with the unknown magnetic field, often adopted in such classes of sensors, has to be taken into account. Therefore, this means that extra features, such as cross-sensitivity and output signal normalized to the auxiliary input, must be characterized in addition to sensitivity, resolution and operating range. Different sensor architectures based on this principle can be developed by using suitable micromachining technologies.

A standard CMOS technology, coupled with a front-side bulk micro-machining post-processing, for the realization of MEMS is considered, for example, for the device reported in Figure 4.1 (Latorre *et al.*, 1998; Baglio *et al.*, 1999b). The large diffusion that standard CMOS technology can ensure in the industrial field to a given product also represents a

Scaling Issues and Design of MEMS   S. Baglio, S. Castorina and N. Savalli
© 2007 John Wiley & Sons, Ltd

Figure 4.1 Scanning electron micrograph of two mechanical structures realized for magnetic field sensing by using anisotropic etching in a standard CMOS process. The smaller structure is named 'D', while the larger is named 'F' (Latorre *et al.*, 1998) (Baglio *et al.*, 1999a). Reproduced with kind permission of SPIE

noticeable feature. Compatibility between CMOS and MEMS technologies means that both mechanical sensing and electrical signal conditioning circuits can be realized on the same IC without any damage to the electronic parts due to the micromachining post-processing.

A basic device may consist of a U-shaped cantilever, as shown in Figure 4.1. The I_f current is driven into the cantilever structure and then the interaction between this current and the magnetic field B produces the Lorentz force, reported in equation (2.10) (chapter 2), where now the L quantity is the length of the cantilever subjected to a perpendicular magnetic field.

The cantilever beam is therefore deflected due to the Lorentz force applied to its free end. Polysilicon strain gauges are embedded into the mechanical structure to detect such deflections, as illustrated in Figure 4.2. The output signal is the V_S voltage across the fixed resistor R that is related to the change dR_G of the polysilicon resistance value R_G. The R_f resistor represents the metal path used to drive the I_f current into the cantilever, while the resistor R_I is used for current-limiting purposes.

The strain gauge resistance variation ΔR_G can be expressed as:

$$\Delta R_G = GR_G\varepsilon \qquad (4.1)$$

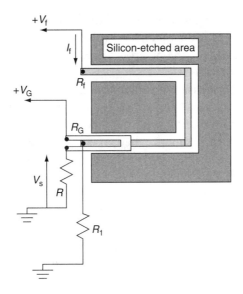

Figure 4.2 Schematic representation of the condensed electromechanical system (Baglio *et al.*, 1999a). Reproduced with kind permission of SPIE

where G represents the gauge factor and the strain ε is linked to the Lorentz force via the stress σ as reported in the following equations:

$$\varepsilon = \frac{\sigma(x, \nu)}{E_n} \tag{4.2a}$$

$$\sigma(x, \nu) = \frac{\nu M(x)}{I_n} \tag{4.2b}$$

$$M(x) = F_L(L_c - x) \tag{4.2c}$$

where E_n and I_n are the Young's modulus and the moment of inertia, respectively. In addition, $M(x)$ is the bending torque applied to the structure in a point at distance x from the clamped end, while L is the whole cantilever length and n represents the distance of the considered point from the neutral axis of the structure. Some simplifying hypotheses have been considered, in particular, a preferential direction for stress has been considered.

A suitable methodology for estimating device characteristics such as gauge factor, Young's modulus and moment of inertia of the mechanical structure has been developed by some of the authors in a previous work (Latorre *et al.*, 1998).

Namely, a static vertical displacement is applied to the mechanical device and both the consequent relative variation of gauges resistance

($\Delta R_G/R_G$) and the strain ε can be estimated. Further efforts can be devoted to the definition of an equivalent Young's modulus E_n and moment of inertia I_n for the heterogeneous structure resulting from the micromachining process.

As specified above, the considered magnetic field sensor is a resonant sensor, in the sense that it requires an auxiliary external input which works in conjunction with the unknown magnetic field in order to produce the output signal. Then, if the forcing current I_f is sinusoidally driven to the resonant frequency of the mechanical structure, the cantilever deflection will be strongly increased by the mechanical resonance of the system.

Cantilever oscillations are hence modulated by the external magnetic field. This allows us to greatly increase the sensitivity of the sensor with respect to the static case. A more complex electronic section is necessary with respect to the static behaviour; however, the output signal level is increased by resonance and a large conditioning circuit gain for suitably detecting the magnetic field is not required anymore, thus lowering the noise effect. Finally, large values are required for the cantilever natural frequency if fast changes in the magnetic field signal

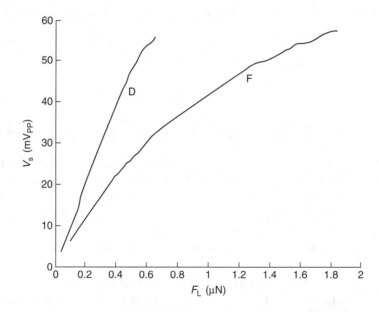

Figure 4.3 Peak-to-peak output voltage for the resonant devices as a function of the applied Lorentz force in the case of constant input (Baglio *et al.*, 1999a). Reproduced with kind permission of SPIE

have to be estimated – this constraint leads to the reduction of cantilever dimensions which is always an appealing condition.

In Figure 4.3 the outputs, i.e. the peak-to-peak amplitude of V_s voltage, of the 'D' and 'F' cantilever are reported for the slowly varying magnetic field in order to perform the device characterization. Both sensitivity and linearity are improved via the sharp reduction in the beam width realized in the 'D' structure.

4.1 INDUCTIVE MICROSENSORS FOR MAGNETIC PARTICLES

As a particular case-study for magnetic sensors, analysis, design and characterization of inductive microsensors will be reported in the following. A discussion on scaling issues will allow drawing interesting conclusions on inductive sensors, despite the simplicity of the structure.

The device structure is a planar differential transformer which could be used for the detection of many kinds of magnetic targets with micrometre or nanometre features sizes. The potential application of the device as a sensor for magnetic immunoassay (Baglio *et al.*, 2002, 2005) is also proposed in this chapter.

4.1.1 Integrated inductive sensors

High-sensitivity magnetic sensors can be designed using different principles of detection, as magnetic resistive transducers, atomic force microscope-based transducers and inductance devices (Webster, 1998). These latter transducers present important advantages, which are related to their higher simplicity, compatibility with standard silicon technology materials, low-cost and higher flexibility. They are based in the detection of changes in the relative magnetic permeability of gas, liquid and solid samples that are positioned inside a measuring coil (Webster, 1998; Ripka, 2001).

Inductive sensors are an interesting 'test-bench' for the analysis of scaling laws, in fact, as thermal sensors and actuators do, they exhibit a simple structure which, in principle, allows integrating them in almost any IC standard technology. Certain applications may require further micromachining processes or the integration of non-standard ferromagnetic materials to improve the performance of the inductor. However, these applications will not be taken into account here.

The analysis that will be presented here is relative to the application of integrated inductive sensors to the detection of small (micro- or nanometres in size) magnetic targets; however, general conclusions can be drawn.

First, a qualitative discussion of the scaling effects on inductive sensors will be given and then the details of the scaling analysis and its applications will be addressed in this chapter.

It can be intuitively understood that the inductance of a given inductor scales with its dimensions. The exact scaling law depends on the structure and geometry of the specific inductor. In principle, the operation of an inductive sensor is based on the detection of the inductance changes induced, for example, by given magnetic targets in the vicinity of the sensors. The sensitivity, expressed as the relative inductance change, is:

$$S = \Delta L / L_0 \tag{4.3}$$

If the inductive sensor is scaled down in size, its inductance diminishes, and from equation (4.3) it follows that a higher sensitivity to small inductance changes may be achieved with scaling. In other words, a scaled inductive sensor can be more sensitive to small magnetic targets, thus showing higher potentials in the detection of micro- or nanomagnetic beads.

On the other hand, small inductance values and even smaller inductance changes are very difficult to detect and thus particular attention must be paid to the design and realization of suitable signal-conditioning circuits, which allow the detection and the processing of the small signals coming from the sensors. From this point of view, the integration of the sensor in standard technologies, like CMOS, is of fundamental importance because it would allow the realization of the signal-conditioning circuit, or at least the most critical part of it, on the same chip of the sensor, thus minimizing most of the inductive and capacitive 'parasitic' effects that would be introduced by external connections, with great benefits for the sensors performances.

In order to focus on an integrated device in standard CMOS technology, it is straightforward to consider planar microcoils. The presence of the magnetic markers (microparticles) in the core of the coil (or near it) produces a change in the inductance value. The detection of this change is therefore a measure of the quantity of particles in the core of the coil, or better stated, of the density of magnetically active material in the core.

Planar inductors can be easily realized in standard CMOS technology and several realizations have been presented in literature for

RF and proximity sensors applications (Chao *et al.*, 2002; Sadler and Ahn, 2001; Ahn and Allen, 1998). However, in order to simplify the readout circuitry and to avoid the coupling and 'parasitic' phenomena, typical of the RF range, working frequencies well below the RF range have been assumed as a design constraint here.

4.1.2 Planar differential transformer

Several approaches to inductive sensing and to the relative signal detection and conditioning have been presented – they are often based on the use of a single coil in filters or oscillators.

A differential approach is used here in order to filter out undesired effects related to interfering signals. A differential transformer where only one of the two secondary windings is made sensitive to the magnetic particles is considered; the other secondary coil allows subtracting spurious effects from the total output of the transformer.

The device presented here is based on a planar coreless differential transformer configuration. A primary coil generates a magnetic flux which links with two secondary coils, with opposite winding sense, connected in a differential arrangement. In Figures 4.4 and 4.5, both the simplified schematics and an illustration of the proposed device are shown.

The primary coil generates a magnetic flux that induces equal but opposite voltages in the secondary coils, due to their opposite winding

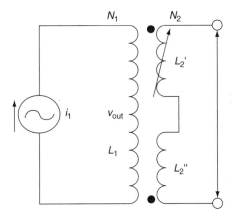

Figure 4.4 Schematic of the principles of the differential transformer (Baglio *et al.*, 2005) © 1996 IEEE. Reproduced by kind permission of IEEE

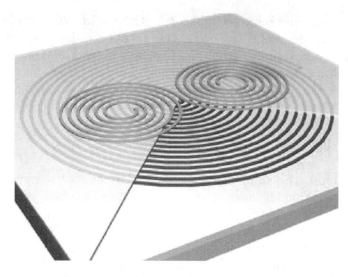

Figure 4.5 Three-dimensional illustration of the planar transformer (Baglio *et al.*, 2005) © 1996 IEEE. Reproduced by kind permission of IEEE

sense; therefore, the resulting output voltage, which is the difference between the voltages across the secondary coils, is zero when no magnetic particles are present. On the other hand, the presence of magnetic particles in one of the secondary coils will cause a redistribution of the magnetic flux lines, which will result denser near the magnetic particles, therefore resulting in a non-zero differential output voltage. This operating principle is schematized in Figure 4.6.

In this approach one of the secondary coils acts as the 'active' sensor, while the other one acts as a 'dummy', as in most differential sensing

$V_{out} = 0$ $V_{out} \neq 0$

Figure 4.6 Working principles of the planar differential transformer (Baglio *et al.*, 2005) © 1996 IEEE. Reproduced by kind permission of IEEE

approaches. In particular, here the differential configuration is used not to enhance sensitivity; in fact, there are no opposite variations of inductance, but to lower the 'noise floor'. The primary coil is a source of excitation of the sensor. This approach allows a more flexible optimization of the device in terms of sensitivity; in fact, in the case of the transformer the open circuit voltage at the secondary coil, if expressed in terms of the current applied to the primary winding, is proportional to the product of the number of turns of the primary and the secondary coils. While the secondary coils can be subjected to more restrictive design constraints due to their sensing function, the primary coils has less restrictions; therefore, both the sensitivity requirements and eventual design constraints can be more easily satisfied with respect to the single inductor case, by proper designing of the primary and secondary coils.

Furthermore, the approach presented here is intrinsically differential, thus allowing a better rejection of noise and interferences. This is suitable for the integration in CMOS technology due to its simple and planar geometry. Moreover, it is not based on the direct estimation of the inductance, resulting in a great simplification of the measurement strategy. In fact, the magnetic particles act as a moveable nucleus and the differential output voltage at the secondary coils is directly related to the number (or density) of magnetic particles. Therefore, a high impedance detection of the differential output voltage at the secondary coils is a simple but good strategy to the detection of the magnetic particles.

The feasibility of this approach has been first validated with 'macro' prototypes realized with printed circuit board (PCB) techniques and iron filing grains instead of the magnetic microparticles, and then CMOS prototypes have been designed, realized and tested with magnetic microparticles.

The simple schematic of Figure 4.4 can be used to describe the operation of the sensor in ideal conditions, but it does not allow taking into account some real operation needs and some 'parasitic' effects. First of all, integrated inductance values are very small and would require operation in the RF field. However, as already explained before, the operation at low frequencies is assumed here as a design constraint; therefore, capacitive loads are connected to the secondary windings in order to limit the resonance frequencies of the circuit.

Furthermore, the small dimensions of the 'tracks' used to realize the coils, especially in the integrated version of the sensor, the absence of a magnetic core and thus the 'non-perfect' magnetic coupling between primary and secondary windings, suggest making use of a circuital model where the non-idealities of the transformer, like the series resistance of the coils, their leakage inductance and their non-ideal coupling, are taken into account. The leakage inductance takes into account for the part of the magnetic flux that does not give a contribution to the magnetic coupling; the magnetizing inductance takes into account the finite inductance of the transformer (Severns and Bloom, 1985; Hui *et al.*, 1999).

Due to the planar structure of the transformer, where the primary winding faces the secondary ones, a capacitive coupling between the windings may affect the operation of the device – therefore, this should be taken into account in the model.

The complete circuit model of the transformer is shown in Figure 4.7, where L_{lki} represents the leakage inductance of the ith coil, L_{m12} and L_{m13} are the magnetizing inductances, R_i is the series resistance of the ith coil, C_{12} and C_{13} are the 'parasitic' capacitances between the primary coil and each secondary coil and R_L and C_L are, respectively, the resistance and the capacitance of the load, which is the signal-conditioning circuit. C_L also takes into account the capacitance used to adjust the resonant frequency. In the following analysis, the current in the secondary windings will be neglected, despite the fact that the open-circuit configuration does not hold for the secondary coils.

The voltage across the load connected to the secondary 2 (and similarly the voltage across the load connected to secondary 3) results in the following:

$$V_2(s) = \frac{sL_{m12}/n}{\dfrac{s^2}{\omega_0^2} + s\left(C_L R_2 + \dfrac{L_{lk2}}{RL}\right) + 1} I_1(s) \qquad (4.4)$$

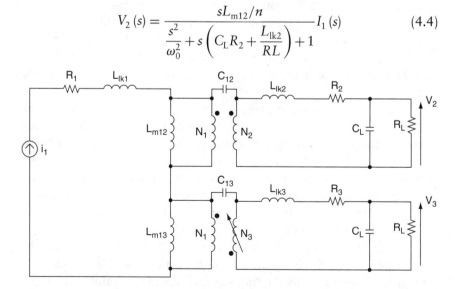

Figure 4.7 Equivalent circuit of the planar transformer (Baglio *et al.*, 2005) © 1996 IEEE. Reproduced by kind permission of IEEE

where:

$$\omega_0^2 = \frac{1}{L_{lk2}C_L} \tag{4.5}$$

and s is the complex frequency.

The gain of the transformer in the frequency domain is:

$$G(s) = \frac{V_2(s) - V_1(s)}{I_1(s)} \tag{4.6}$$

The circuit parameters that appear in Figure 4.8 are given by:

$$n = \frac{N_1}{N_2} = \frac{N_1}{N_3} \tag{4.7}$$

$$L_{m12} = nM_{12} \tag{4.8}$$

$$L_{lk2} = L_2 - \frac{L_{m12}}{n^2} \tag{4.9}$$

If the variable inductance is expressed as $L_3 = L_2(1+x)$, with $x = \Delta L/L_0$, thus the mutual inductance results in $M_{13} = M_{12}(1+x)^{1/2}$, which gives:

$$L_{m13} = nM_{12}\sqrt{1+x} \tag{4.10}$$

$$L_{lk3} = L_{20}(1+x) - \frac{M_{12}}{n}\sqrt{1+x} \tag{4.11}$$

$$L_{lk1} = L_1 - \frac{L_{m12} + L_{m13}}{n^2} \tag{4.12}$$

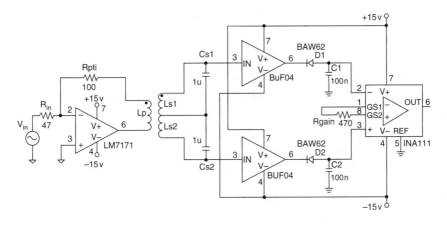

Figure 4.8 Signal-conditioning circuit for the planar differential transformer (Baglio *et al.*, 2005) © 1996 IEEE. Reproduced by kind permission of IEEE

The coupling capacitance between the primary coil and each of the secondary ones has been calculated by approximating the coils to circular plates. This is a worst-case approximation and therefore it produces a 'conservative' value. This value is low enough to have no appreciable effects at the operating frequencies of the device.

The self and mutual inductances of the coils have been calculated by using the method proposed by Hurley et al., (1999), which proposes new formulae for the self- and mutual-impedance calculation of planar coils.

The inductive parameters of planar coils or transformers depend on the geometry. In Hui et al. (1999), an extensive analysis of the effects of geometrical parameters on the self- and mutual-inductances is presented, together with the experimental verification of the results. Some interesting considerations on the scaling of planar coils come from such analysis. In particular, the coil outermost radius and the number of turns affect all of the inductive parameters significantly. The self-inductance of a coreless planar transformer is a linear function of the transformer outermost radius. The mutual and leakage inductances are also linear functions of the outermost radius, provided that it is much greater than the dielectric thickness. The inductive parameters are second-order functions of the number of turns, when the outermost radius is fixed. In the case of fixed-track separation, the inductive parameters are third-order functions of the number of turns. The thicker the dielectric layer, the smaller the mutual inductance becomes. However, the self-inductance is not affected by the dielectric thickness significantly. The conductor width and thickness do not heavily affect the inductive parameters.

4.1.3   Signal-conditioning circuits

The signal-conditioning circuit has been implemented in a discrete form by using commonly available electronic components and integrated circuits. It consists of a voltagetocurrent converter which supply a constant-amplitude sinusoidal current to the primary winding. This allows supplying an excitation current which is insensitive to the impedance of the primary coil to a certain degree. The difference in the peak voltages at the secondary coils at a given operating frequency is of interest in the proposed application; therefore, the signal-conditioning circuit at the secondary side is composed of two peak detectors, which are connected to an instrumentation amplifier through two unity gain buffers in order to perform high-impedance reading. Load capacitances are connected in parallel to the secondary coils to reduce the resonance frequency value. The complete circuit diagram of the signal conditioning circuit is shown in Figure 4.8.

4.1.4 Simulation of the planar differential transformer

The equations developed by Hurley et al., (1999) have been implemented in 'MATLAB' and used to simulate the behavior of the device. The simulated gain, defined as the differential output voltage at the secondary coils, dividing the current in the primary, for a relative inductance change from 0 % to 2.5 %, is shown in Figure 4.9 as a function of frequency, and in Figure 4.10 as a function of the relative inductance change. Such a simulation is relative to the transformer with the geometrical and electrical features summarized in Table 4.1.

The plots in Figure 4.9 and Figure 4.10 show, as expected, that the change in the inductance value produces a variation in the gain value, with a maximum around the resonance frequency. Furthermore,

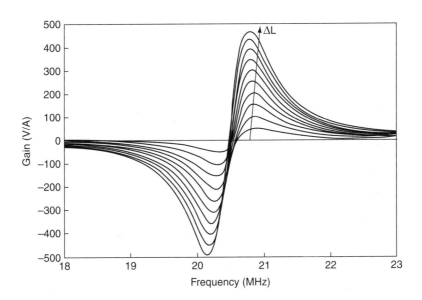

Figure 4.9 Simulated gain of the transformer for relative inductance changes between 0 % and 2.5 % (step, 0.25 %) (Baglio et al., 2005) © 1996 IEEE. Reproduced by kind permission of IEEE

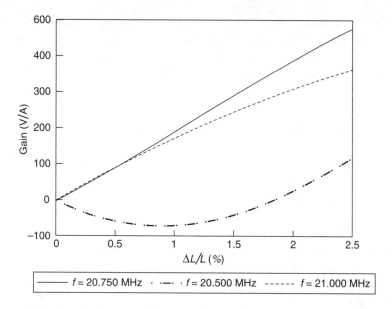

Figure 4.10 Simulated gain of the transformer versus relative inductance change at different frequencies (Baglio *et al.*, 2005) © 1996 IEEE. Reproduced by kind permission of IEEE

Table 4.1 Geometrical and electrical features of a transformer (Baglio *et al.*, 2005) © 1996 IEEE. Reproduced by kind permission of IEEE

Parameter	Primary coil		Single secondary coil
Number of turns	30		5
External diameter (mm)	40		10
Internal diameter (mm)	0		4
Track height (μm)		35	
Track width (mm)		0.2	
Separation between tracks (mm)	0.46		0.50
Separation between coils (mm)		0.50	
Inductance (nH)	8020.2		199.34
Series DC resistance (Ω)	5.4		0.3
Mutual inductance (nH)		443.21	
Coupling factor		0.35	
Coupling capacitance (fF)		1.96	

the variation of the gain with the relative inductance change is linear with good approximation for small values of ΔL. The change in the transformer gain value is directly correlated to a change in the differential output voltage, if a constant-amplitude sinusoidal current is provided to the primary winding.

4.1.5   Experimental results

The proposed device and measurement strategy have been experimentally verified at two different dimensional scales by means of a macroprototype, realized in printed circuit board (PCB) technology, and a microprototype, integrated in a standard CMOS technology. In such a way, the effects of scaling on the effectiveness of the device and the measurement strategy have been validated.

For the PCB device, the primary and secondary windings have been realized on separate boards and then superimposed. The layout of the windings and a picture of the two boards are shown in Figure 4.11. Here, circular windings have been used in order to minimize the series resistance and the 'parasitic' capacitance, thanks to the lower areato-perimeter ratio.

The PCB prototype shown in Figure 4.11 has been characterized by using iron filing grains as magnetic beads to be detected. The grains have been placed at the centre of a secondary coil and the corresponding output voltage has been measured. The experimental setup and procedure are shown in Figure 4.12.

In a first-characterization procedure, the iron filings have been placed in the PCB transformer core in quantities corresponding to integer multiples of a well-reproducible one; the weight of such a 'unitary' amount of iron filings has been measured with an assay balance, and its value is $133.6 \pm 0.1 \, \mu g$. The amounts of iron filings used in this phase goes from one to four times the 'unitary' quantity. In Figure 4.13, the measured transformer gain is shown, together with the simulated plots that better

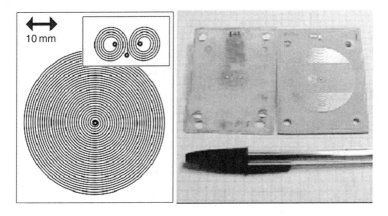

Figure 4.11   The PCB transformer prototype (Baglio *et al.*, 2005) © 1996 IEEE. Reproduced by kind permission of IEEE

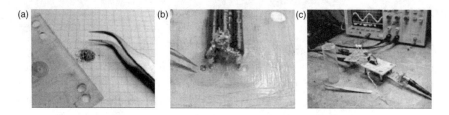

Figure 4.12 Placement of the iron filings on the PCB prototype (Baglio *et al.*, 2005) © 1996 IEEE. Reproduced by kind permission of IEEE

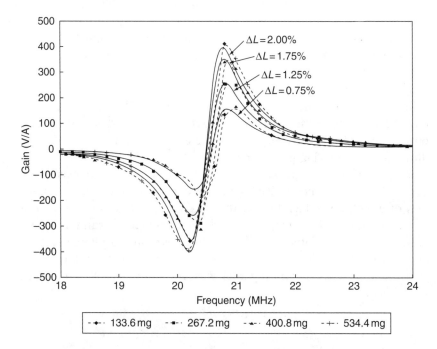

Figure 4.13 Measured gain of the transformer for different amounts of iron filings in a secondary coil core (dotted lines with markers). The best-simulated plots are also reported for comparison (continuous lines). The relative inductance variation caused by a given amount of iron filings has been approximately estimated by comparing simulated and experimental results (Baglio *et al.*, 2005) © 1996 IEEE. Reproduced by kind permission of IEEE

fit the measured data. In such a way, the inductance variation induced by a given amount of iron filings has also been approximately estimated. The measured differential output voltage at 21 MHz, i.e. close to the resonance peak, for different amounts of iron filings is reported in Figure 4.14. From this latter plot, an average sensitivity to the presence of iron filings of 2.6 mV/mg can be estimated.

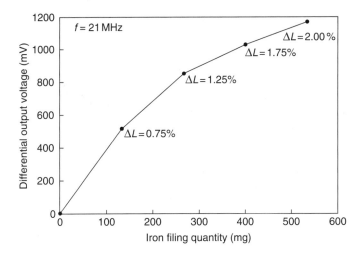

Figure 4.14 Measured differential output voltage of the transformer versus quantity of iron filings. The average sensitivity results in a value of 2.6 mV/mg ($f = 21$ MHz) (Baglio *et al.*, 2005) © 1996 IEEE. Reproduced by kind permission of IEEE

The operating frequency can be reduced by increasing the load capacitance. However, a significant decrease in the output voltage and, then, in the sensitivity has been predicted and observed; therefore, a 'trade-off' should be accurately chosen.

In a second phase of the characterization, a more refined validation of the device has been performed by placing single grains of iron filings. The results relative to the differential output voltage at 9.7 MHz for 100 grains are shown in Figure 4.15.

Two different series of measurements are reported. From these results, good linearity and reproducibility arise. The differences between the two plots and the 'steps' in their trend could be due to the non-homogeneities in the grain size. From Figure 4.15, a qualitative sensitivity of about 1 mV/grain can be estimated.

Now, the design, simulation and characterization of the CMOS microprototype will be reported. The technology of choice is the 0.8 μm 'CMOS CXQ' process by Austria Mikro Systeme (AMS). This is a standard CMOS technology with two metal layers. These layers have been used to realize the windings of the planar transformer. The primary winding has been realized in the metal 1 layer, while the metal 2 has been used to realize the secondary windings. The secondary coils are separated from the primary one by the VIA oxide. The passivation layer covers the whole transformer, except for the central area of the secondary coils.

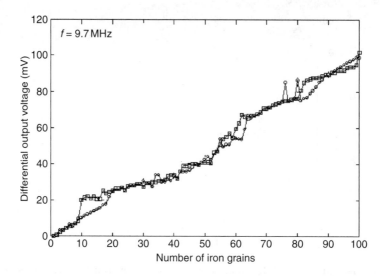

Figure 4.15 Experimental 'single-grain' characterization of the planar transformer ($f = 9.7\,\mathrm{MHz}$) (Baglio *et al.*, 2005) © 1996 IEEE. Reproduced by kind permission of IEEE

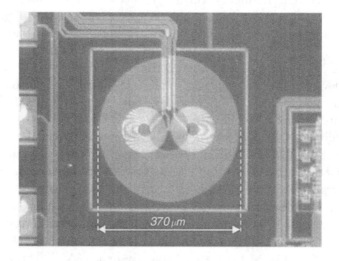

Figure 4.16 Micro-photography of the circular integrated transformer prototype (Baglio *et al.*, 2005) © 1996 IEEE. Reproduced by kind permission of IEEE

An illustration of a circular transformer prototype is reported in Figure 4.16, while a schematic cross-section is shown in Figure 4.17. The geometrical and electrical parameters of such a microtransformer are summarized in Table 4.2, where the self- and mutual-inductance values

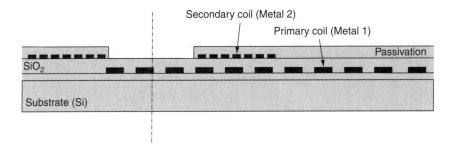

Figure 4.17 Schematic of the cross-section along a radial direction (Baglio *et al.*, 2005) © 1996 IEEE. Reproduced by kind permission of IEEE

Table 4.2 Geometrical and electrical features of the CMOS microtransformer (Baglio *et al.*, 2005) © 1996 IEEE. Reproduced by kind permission of IEEE

Parameter	Primary coil		Single secondary coil
Number of turns	26		10
External diameter (μm)	370		116
Internal diameter (μm)	110		36
Track height (μm)	0.6		1
Track width (μm)	3		2
Separation between tracks (μm)	2		2
Separation between coils (μm)		0.8	
Inductance (nH)	157		7.1
Series DC resistance (Ω)	458		67.5
Mutual inductance (nH)		8.4	
Coupling factor		0.35	
Coupling capacitance (pF)		0.46	

have been calculated with the method proposed by Hurley *et al.* (1999). The coupling capacitance between the primary and each secondary has been calculated by considering the spiral windings as circular plates.

A characterization of the transformer without any capacitive load connected to the secondary coils has been performed. Such a characterization is important to evaluate the presence and entity of offsets due to eventual misalignment between the primary and secondary coils.

In fact, despite the PCB macroprototype, in the integrated device the misalignment cannot be corrected mechanically. The results of these measurements are reported in Figure 4.18. It can be highlighted that, without any capacitive load, the self-resonant frequency of the system made by the primary and one secondary coil is about 34 MHz, too high for the scope of the work presented here. Therefore, as for the PCB prototypes, load capacitance will be connected in parallel to each secondary coil to reduce the resonant frequency to a convenient value.

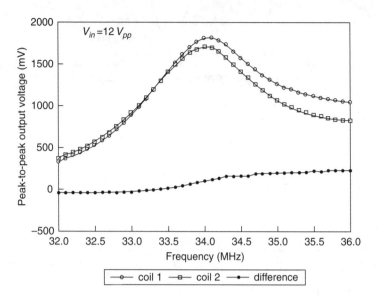

Figure 4.18 Measured frequency response of the CMOS integrated microtransformer with open-circuit secondary coils ($V_{in} = 12V_{pp}$) (Baglio *et al.*, 2005) © 1996 IEEE. Reproduced by kind permission of IEEE

The circuit shown in Figure 4.8, in the case of the CMOS microtransformer, has been assembled on a suitably designed PCB. In the realization of such a circuit and PCB, particular attention has been devoted to shielding and minimizing electro-magnetic interferences. The whole PCB has been placed in a metallic shield box and connected to the measurement instrumentation through coaxial shielded cables. Two illustrations of the assembled system are reported in Figure 4.19.

The characterization of the device has been performed by placing over one secondary coil increasing quantities of micromagnetic particles and by detecting the amplified difference in the peak voltages at the secondary coil over a wide range of frequencies. The particles used for the characterization were 'SPHERO^{TM} Polystyrene Carboxyl Magnetic Particles, Smooth Surface', by Spherotech, Inc. A 0.5 ml 2.5 % w/v sample of particles with an average diameter of 4.1 μm, containing a total amount of 12.5 mg of particles having a density of 1.05 g/cm^3, has been used for the characterization.

Since only one sample of magnetic particles with a given concentration was available for the device characterization, the increase of concentration has been 'simulated' by incremental deposition of the sample over the sensors. First, a given amount of sample volume was withdrawn, placed over the sensor, dried out and then the output voltage was

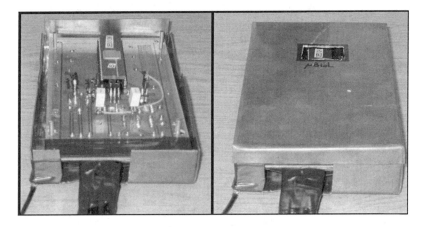

Figure 4.19 Experimental assembly of the CMOS microtransformer and its signal-conditioning electronics (Baglio *et al.*, 2005) © 1996 IEEE. Reproduced by kind permission of IEEE

measured. The successive characterization steps have been performed by adding other amounts of sample to the dry quantity already deposited over the sensor, in such a way that several concentrations have been simulated. The lack of a repeatable withdrawal system did not allowed us to get the same amount of sample each time; however, this does not affect the validity of the characterization method adopted.

Four different amounts of sample have been withdrawn in terms of percentage of the initial sample volume. The samples have been agitated before each withdrawal to ensure a homogeneous distribution of particles.

The four phases of the characterization, with the dried particles deposited over the sensor, are show in Figure 4.20. The results of the characterization are reported in Figure 4.21, in terms of the peak voltage difference between the secondary coils, in the frequency range 10–18 MHz, while in Figure 4.22 the response of the system in terms of output voltage versus sample concentration at 13.2 MHz is reported.

From the plot of Figure 4.22, a sensitivity of 2.93 % V_{out}/mg/ml can be estimated for concentrations higher than 25 mg/ml.

To compare the responses of both the device prototypes presented here, the PCB and the CMOS ones, it is necessary to express the response of the integrated transformer in terms of particles weight instead of concentration; in fact, the characterization of the PCB transformer has been done by using 'dry' deposited iron filings. The responses of the two devices are compared in Figures 4.23 and 4.24, in logarithmic and normalized scales, respectively.

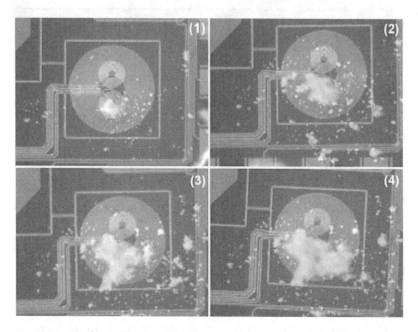

Figure 4.20 The four phases of the CMOS microtransformer characterization by using magnetic particles (Baglio *et al.*, 2005) © 1996 IEEE. Reproduced by kind permission of IEEE

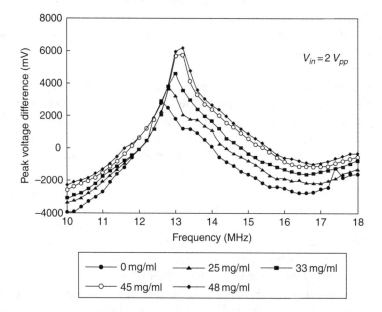

Figure 4.21 Experimental results obtained for the characterization with magnetic particles ($V_{in} = 2V_{pp}$) (Baglio *et al.*, 2005) © 1996 IEEE. Reproduced by kind permission of IEEE

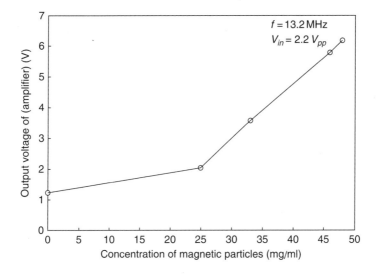

Figure 4.22   Characterization of the sensor with magnetic particles. The amplifier's output voltage is reported versus different 'simulated' concentration values of the sample ($f = 13.2\,\text{MHz}$; $V_{in} = 2.2V_{pp}$) (Baglio *et al.*, 2005) © 1996 IEEE. Reproduced by kind permission of IEEE

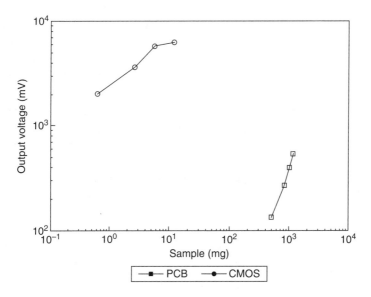

Figure 4.23   Comparison between the responses of the PCB and CMOS transfer prototypes in terms of sample weight (Baglio *et al.*, 2005) © 1996 IEEE. Reproduced by kind permission of IEEE

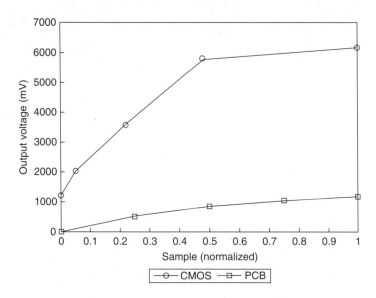

Figure 4.24   Comparison between the responses of the PCB and the CMOS transformer prototypes in terms of sample weight normalized to the maximum sample amount for each case (Baglio *et al.*, 2005) © 1996 IEEE. Reproduced by kind permission of IEEE

Both Figures 4.23 and 4.24 show the improvement in sensitivity achieved with the integrated sensor; in fact, the output voltage of the circuit employing the integrated device is higher than the PCB one and so a high output voltage is achieved with smaller amounts of sample. Such a gain in terms of sensitivity is even higher if the different nature of the samples used for the characterization of the two prototypes is considered. In fact, the paramagnetic particles used for the characterization of the integrated sensor have a relative magnetic permeability at least one order of magnitude smaller than those of the ferromagnetic iron filings used in the case of the PCB device.

In terms of sample weight, the sensitivity of the integrated prototype can be estimated as 730 mV/mg, which is 280 times higher than the sensitivity of the PCB one.

As a further comparison between the two versions of these devices, the PCB device shows a negligible or null offset, compared to the CMOS one; this is due to the physical realization of the PCB transformer, based on two overlapped printed circuit boards, whose alignment can be adjusted and thus the offset resulting from coils misalignment can be compensated. Of course, this is not possible in the CMOS realization. Furthermore, many more parameters affect the behaviour of the integrated device than the PCB one. However, an offset in the device response can be compensated for by a suitable signal-conditioning circuit – therefore, it does not represent a problem.

4.2 MAGNETIC IMMUNOASSAY SYSTEMS

There is an important need for the development, in the next years, of low-cost and high-performance transducers for the detection of biological agents. This is critical in many fields related to public health, the food industry, water management and clinical and diagnostic analysis. In particular, requirements on public health and environmental impact demand for the availability of high-sensitivity, low-cost and simple analytical tools. All of this imposes the need of sensors, for application to biological systems, combining the characteristics of low-cost, high-sensitivity and specificity, with short analysis times, ease to handle and ease to transport for *insitu* and in-field measurements.

High sensitivity and specificity can be obtained by using immunological techniques (Larsson *et al.*, 1999) which are based on the biological recognition of the analyte to be detected by specific antigens or antibodies. Detection is made by coupling these molecules to suitable markers, such as radioactive compounds, enzymes, fluorophores and luminescent ones. In front of these, magnetic markers have potential advantages, which are related to their low price, very high stability and absence of toxicity. In addition, biomolecules fixed to magnetic nanoparticles can be easily localized and manipulated by suitable magnetic fields.

The problem of detecting biological agents is therefore shifted to the ability of sensing the presence of a small number of magnetic particles.

On the other hand, the requirements related to low-cost, ease of handling and portability demand the development of integrated micro-machined devices in a Micro Total-Analysis System (μTAS). This will benefit not only the cost reduction inherent to mass production of Si-based micromachining technologies, but also the possibility for the design of more complex low-cost devices and sensor arrays.

The application of the measurement system proposed here to magnetic immunoassays requires the realization of functionalized surfaces, the coil core or its whole surface (Edelstein *et al.*, 2000) and the magnetic particles to be used as markers. Such functionalization consists of the coating of the sensors and particles surfaces with suitable materials to allow the binding of suitable antibodies, as schematized in Figure 4.25(a)

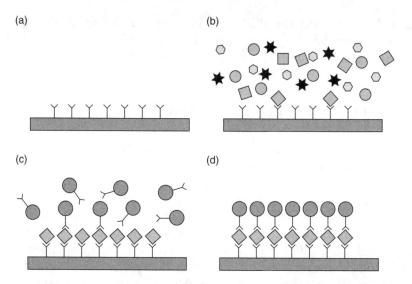

Figure 4.25 Schematics of detection principles: (a) the sensor's surface is functionalized with the specific antibodies; (b) only the specific analyte in the sample binds to its antibody on the surface; (c) the functionalized markers bind to the trapped analyte; (d) the 'sandwich' is ready for detection

and Figure 4.25(c). These steps are essential to apply a given detection system or method to immunoassay and to achieve the required specificity.

As shown in Figure 4.25(a)–4.25(d), only the targeted analyte binds to its specific antibodies, remaining trapped to the sensor surface; then, the functionalized markers bind to the trapped analyte which can be detected. Once the 'sandwich' represented in Figure 4.25(d) is formed, the problem is to detect the number or density of the markers. Suitable methods and systems should be adopted, depending on the type of markers used. From this point of view, the proposed inductive microsensors have promising characteristics to be applied in future lowcost, portable high-sensitivity magnetic immunoassay devices.

In the specific case of magnetic immunoassays based on inductive sensors, the quantitative detection of the analyte in the specimen is determined by the amount of magnetic beads fixed to the surface, which in turn, determines a change in the inductance of the coil. Measurement of the inductance with a simple electronic circuit allows quantification of the analyte content in the specimen.

The sensitivity shown by the integrated microtransformer prototype, $2.93\,\%V_{out/mg/ml}$, results in being smaller than the several relative fluorescence units/μg/ml reported for some fluorescence immunoassay methods (Kamyshny and Magdassi, 2000; Sun *et al.*, 2001), but it may represent a first step toward the development of integrated, lowcost, portable magnetic immunoassay systems.

With reference to the application of the inductive magnetic biosensors presented here to the field of immunoassay, a potential use of such sensors in a magnetic 'micro-Total-Analysis-System' (μ-TAS) has been devised. The envisioned system could be made of several basic microfluidic components, such as microchannels, microvalves, micropumps, flow sensors, microreservoirs and microneedles. A schematic of the principles of such a system is provided in Figure 4.26.

The operation of the proposed magnetic μ-TAS is the following: the fluid to be analysed is taken from the external environment through a microneedle by means of a micropump; a microvalve allows the fluid flowing only or mainly in the right (inward) direction; flow sensors can be used to determine the amount of fluid withdrawn; the fluid comes in to a reservoir, the detection chamber, having a surface suitably treated with the specific antibodies; the fluid containing the magnetic markers, stored in a separated reservoir, is then injected in the detection chamber where the immunoassay sandwich (antibody/analyte/antibody with magnetic markers) is therefore formed; excess and exhaust fluids will be ejected out from the detection chamber; at this point, the trapped magnetic particles, and then the analyte can be detected by the inductive sensor; the sensor output signal is processed by the on-board electronics.

Arrays of such magnetic μTASs could be realized on the same chip and each section of the array could be made sensitive to different analytes, which could be detected in parallel.

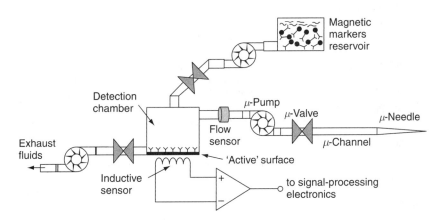

Figure 4.26 Schematic of a potential μ-TAS based on the inductive magnetic biosensors presented here

Different kind of sensors could be integrated on the same substrate of the proposed magnetic μTAS, such as temperature sensors, pressure sensors, humidity sensors and even gas sensors, thus realizing a complete environmental monitoring or sampling system on a chip suitable for distributed sensing.

ACKNOWLEDGEMENTS

1. 'Magnetic field sensors are widely apopted . . . in the beam width realized in the D structure.', pp. 73–77. Portions of the text are reproduced from S. Baglio, L. Latorre and P. Nouet (1999a). *Development of Novel Magnetic Field Monolithic Sensors with Standard CMOS Compatible MEMS Technology*, SPIE'99, Vol. 3668, Los Angeles, pp. 417–424 and are reproduced by kind permission of SPIE.
2. Sections 4.1–4.2 'The device structure is a planar . . . a chip suitable for distributed sensing.', pp. 77–100. Portions of text are reproduced from S. Baglio, S. Castorina and N. Savalli (2005) Integrated inductive sensors for magnetic immunoassay applications, *IEEE Sensors Journal*, 5 (3), 372–384, © 1996 IEEE and are reproduced by kind permission of IEEE.

REFERENCES

C.H. Ahn and M.G. Allen (1998). Micromachined planar inductors on silicon wafers for MEMS applications, *IEEE Trans. Ind. Elect.*, **45**, 866–876.

S. Baglio, L. Latorre and P. Nouet (1999a). *Development of Novel Magnetic Field Monolithic Sensors with Standard CMOS Compatible MEMS Technology*, SPIE'99, Vol. 3668, Los Angeles, pp. 417–424.

S. Baglio, L. Latorre and P. Nouet (1999b). Resonant magnetic field microsensors in standard CMOS technology, in *IMTC/99, Proceedings of the 16th IEEE Instrument and Measuring Technology Conference*, Vol. 1, Venice, Italy, May 24–26, pp. 452–457.

S. Baglio, S. Castorina and N. Savalli (2002). Strategies and circuits for micro inductive bio-sensors conditioning, in *Proceedings of the 16th European Conference on Solid-State Transducers (EUROSENSORS XVI)*, Prague, Czech Republic, September 15–18, pp. 604–607.

S. Baglio, S. Castorina and N. Savalli (2005) Integrated inductive sensors for magnetic immunoassay applications, *IEEE Sensors Journal*, 5, 372–384.

C.J. Chao, S.C. Wong, C.H. Kao, M.J. Chen, L.Y. Leu and K.Y. Chiu (2002). Characterization and modeling of on-chip spiral inductors for Si RFICs, *IEEE Trans. Semiconductors Manuf.*, **15**, 19–29.

R.L. Edelstein, C.R. Tamanaha, P.E. Sheehan, M.M. Miller, D.R. Baselt, L.J. Whitman and R.J. Colton (2000). The BARC biosensor applied to the detection of biological warfare agents, *Biosensors Bioelectron.*, **14**, 805–813.

J.L. Gardner (1994). *Microsensors, Principles and Applications*, John Wiley & Sons, Chichester.

S.Y. Hui, H.S. Chung and S.C. Tang (1999). Coreless printed circuit board (PCB) transformers for power MOSFET/IGBT gate drive circuits, *IEEE Trans. Power Elect.*, **14**, 422–430.

W.G. Hurley, M.C. Duffy, S. O'Reilly and S.C. O'Mathúna (1999). Impedance formulae for planar magnetic structures with spiral windings, *IEEE Trans. Ind. Elect.*, **46**, 271–278.

Z. Kadar, A. Bossche and J. Mollinger (1994). Integrated resonant magnetic-field sensor, *Sensors Actuators A: phys.*, **70**, 225–232.

A. Kamyshny and S. Magdassi (2000). Fluorescence immunoassay based on fluorescer micro-particles, *Colloids Surfaces B: Biointerfaces*, **18**, 13–17.

L. Latorre, Y. Bertrand and P. Nouet (1998). On the use of test structures for the electromechanical characterization of a CMOS compatible MEMS technology, in *Proceedings of the IEEE 1998 International Conference on Microelectronic Test Structures*, Vol. 11, Kanazawa, Japan, March 23–26, pp. 177–182.

K. Larsson, K. Kriz and D. Kriz (1999). Magnetic Transducers in Biosensors and Bioassays, *Analusis*, **27**, 617–621.

P. Ripka (Ed.) (2001). *Magnetic Sensors and Magnetometers*, Artech House, Norwood, MA.

C.S. Roumenin (1994). *Solid State Magnetic Sensors*, Elsevier, Amsterdam, The Netherlands.

D.J. Sadler and C.H. Ahn (2001). On chip eddy current for proximity sensing and crack detection, *Sensors Actuators A: Phys.*, **91**, 346–351.

R.P. Severns and G.E. Bloom (1985). *Modern DC-to-DC Switchmode Power Converter Circuits*, Van Nostrand Reinhold, New York, NY, USA.

B. Sun, W. Xie, G. Yi, D. Chen, Y. Zhou and J. Cheng (2001). Microminiaturized immuoassays using quantum dots as fluorescent label by laser confocal scanning fluorescence detection, *J. Immunol. Meth.*, **249**, 85–89.

J.G. Webster (Ed.) (1998). *The Measurement, Instrumentation and Sensors Handbook*, Electrical Engineering Handbook Series, CRC Press, Boca Raton, FL, USA.

5

Scaling of Mechanical Sensors

5.1 INTRODUCTION

Micromechanical sensors can be roughly defined as sensors which are sensitive in one way or another to mechanical quantities and which are made by micromachining technologies.

Almost all sensors providing information about mechanical variables have gained from the reduction of the scale size of microelectromechanical structures. Among them, scaling of inertial sensors (often referred to for measuring spatial variables, such as displacement, acceleration or angular rate, etc.) and mass sensors (referred to as mechanical sensors in Webster (1999)) have been widely investigated and produced in large quantities. Accelerometers, in particular, have actually the second largest sales volume after pressure sensors and gas sensors. Miniaturization has opened upto such a class of microdevices a wide applications scenario due to the extension of the classical operating ranges and improvements in resolutions and sensitivity; in addition, giving the possibility to perform distributed measurements in hostile or at least special environments, as in the cases of spatial aircraft and biomedical applications.

A huge amount of work can be found in the literature dealing with many microtechnologies which propose to optimize several design aspects and to satisfy different requirements for a wide scenario of application field (Yazdi *et al.*, 1998; Kraft, 2000). Most of them concentrate their attention to improving aspects inherent to the signal output

Scaling Issues and Design of MEMS   S. Baglio, S. Castorina and N. Savalli
© 2007 John Wiley & Sons, Ltd

conditioning strategy while others focus losses minimization or issues arising during devices fabrication. It is evident that a vast category of technical and scientific competencies move around this relatively new world, while market trends have just identified winning products and depicted future needs.

Apart form this, it is clear that scaling issues in designing structures often aiming or looking at mesoscale or nanoscale counterparts must be taken into account from the conception of a new device in a specific field. They play an important role, and define semi-empirical design rules, both during the fabrication process and the device operating life.

Powerful simulation tools give rise to the opportunity to estimate the operating conditions that such devices will experience since the photolithographic steps of the focused technology has been completed. Moreover, analytical or numeric supports are not always sufficiently efficient in predicting effects that micromachining procedures will really have on device integrity or functionality. All micromachining techniques, in fact, induce variations in both the mechanical or electrical, and hence physical, properties of materials of the selected process, depending prevalently on the chemical composition of materials. Scaling of materials properties must be carefully taken into account since many disparities arise in the process used to produce thin-film materials with respect to bulk materials. Important material properties include elastic modulus, Poisson's ratio, fracture stress, yield stress, residual in-plane stress, vertical stress gradient, resistivity, etc. Moreover, it is worthwhile to highlight for the purpose of this book that assumptions such as that of homogeneity become unreliable when used to model device that have dimensions on the same scale of the material's local defects density (Judy, 2001).

In addition, due to the high surface-to-volume ratio of micromechanical sensors, more attention must be paid to controlling their surface characteristics. In fact, scaling down a 'macrosensor' from 1 cm to $10\,\mu m$ means that the surface forces will play a 1000-fold more important role. This is, for example, reflected in the fact that gravity normally does not play an important role in micromechanical sensors (nor do inertial forces; as an example, the deflection of a beam under its own weight scales as l^2 – so a ten times smaller beam bends 100 times less), except when sensors are especially aimed at measuring these effects; in fact, in these situations the small mass associated with micromechanical sensors is a disadvantage.

Chemical composition and consequent action of typical wet or plasma etchants onto the material's properties must be unavoidably considered.

Relevant forces, arising during moveable structures releasing, will not be endured in the same way by all device parts, depending on their dimensions. Since such 'releasing stress' is proportional to the exposed surfaces it will affect the final result depending on structure compliance.

5.2 DEVICE MODELLING AND FABRICATION PROCESSES

Shrinking dimensions hence give the possibility to optimize performances of the devices focused on in this chapter, more specifically accelerometers and resonant mass sensors. Some trivial calculations shown as scaling of linear dimension in a reference device will allow us to make the right choice and adopt it for a specific purpose rather than for any other.

This chapter will hence give a description of the aforementioned effects of scaling, related to the adopted fabrication process, on operating performances of microaccelerometers and masssensors referring to two of the most common micromachining techniques and microtechnologies. More than for devices discussed in other chapters, it is not possible to go into detail for all of the specifics of micromechanical sensors concerns, simply since we neither have the room nor the time to cover an area of research being so wide and so multidisciplinary in nature.

5.2.1 Fabrication processes

Current techniques adopted to release mechanical elements in MEMS include two main approaches, consisting of *surface micromachining* (SM) and *front-side or back-side bulk micromachining*, addressed as FSBM and BSBM, respectively (Madou, 2002; Kovacs, 1998).

Briefly, in the case of surface micromachining, the micromechanical structures are fabricated by processing the surface of the wafer, using a combination of structural, 'sacrificial' and etch-stop layers deposited or epitaxially grown. This process involves two main technological steps: the first step consists of highly anisotropic vertical dry etching in order to define the structural layers; the second step involves surface micromachining based on a selective wet lateral 'under-etching' of the sacrificial layers. The maximal dimensions of the released structures are dependent on the final drying procedure. After wet 'sacrificial' layer etching, the released structure is immersed in the liquid. During the subsequent drying, structures can be pinned down to the substrate by capillary forces, thus inducing 'stiction'. It is important to minimize adhesion forces, in this case, either by reducing the contact area or by changing the surface properties (Tas *et al.*, 1996), since surface roughness plays an important role in adhesion. Some specific drying procedures have been developed by many authors to prevent problems with sticking during release and several methods propose to apply temporary supports to counteract the surface tension forces during drying. Polysilicon supports are used but these have to be removed after release-processing (Fedder *et al.*, 2000; Guillou *et al.*, 2000). Therefore, stiction is a big issue to be taken into account during device design which will experience surface micromachining. It is clear that stiction can easily cause malfunctioning in many devices. The critical length depends less than proportionally on the thickness, gap spacing and adhesion energy.

Bulk micromachining usually refers to etching through from the backside, where time-stop etching, a so-called p^+ etch-stop layer, or a suitable buried oxide layer, can be used to define the structure's geometry. As the cost of the post-process is strongly affected by the needs for alignment, FSBM seems also to be a very promising technique due to its self-alignment capability. In this technique, the post-process operates as an anisotropic etching of the silicon substrate that, i.e. at the end of the process, is exposed in some suitable regions, and uses different layers as masks depending on the selected technology. Suspended structures, such as cantilever beams, bridges or membranes are then obtained. Each structure consists in a heterogeneous stack of the various layers of the process, namely silicon oxides, metals, polysilicon and passivation nitride. The metalization and dielectric layers, normally used for electrical interconnections, now serve a dual function as structural layers. However, there exist material limitations and constraints. Most notable is the large vertical residual stress gradient due to the composite nature of the structures, which cause 'curling out' of the plane of the substrate and which may vary from run to run, often inducing the structures to collapse.

It is clear that issues related to micromachining procedures strongly influence device design scaling potentials, independently from the adopted strategy. Etchant solutions or process materials will influence device operating conditions and integrity, whereas conservative design conditions do not allow us to optimize products' performances.

5.2.2   Devices modelling

Regarding the realization of accelerometers and mass sensors, several alternative solutions have been proposed during the last two decades. The requirements for inertial sensors vary drastically for the applications that they are intended to be used for. These sensors can be, in fact, used in a wide range of applications and have a number of significant advantages, similar to those of integrated circuits, over their conventional counterparts. They are cheaper, since they can be batch-fabricated, can be fitted into a small volume, have smaller form factors and lower power consumption, suitable for battery-operated devices.

Both of them generally consist of a 'proof-mass' suspended by compliant beams anchored to a fixed frame or simply by a cantilever, as mentioned in Chapter 1. A linear restoring coefficient and damping factor can be assumed for analytical models that are usually developed during the preliminary design phase. External acceleration will induce relative proof-mass displacement with respect to the fixed frame, thus resulting in changes of the internal stress in suspension springs. Then, several different readout strategies have been developed since both the relative mass displacement and the suspension-springs stress can be used as a measure of the external acceleration.

For improving understanding of scaling laws in such classes of sensors, almost never intuitive nor simply to transfer from the macroworld to the microdomain, consider a reference device made of a given material, or a stack of different materials, having dimensions of the square proof-mass side L, and thickness t_M, supported by compliant beams of the same material, having dimensions l, w and t_B, for length, width and thickness, respectively. The residual stress and extensional stress are neglected in the following analysis.

Given ρ as the density of the material, the mass of the cantilever is:

$$M = \rho(L^2 t_M + 4lwt_B) \tag{5.1}$$

The elastic constant of the structure can be assumed as being four times the beam's spring constant along the z-direction and derived from equation (5.2):

$$K_z = 48\frac{YI}{L^3} \tag{5.2}$$

where the product YI is the flexural rigidity of the beam, Y is the material's Young's modulus and I is the cross-sectional momentum of inertia, which is proportional to wt^3. Moreover, for analytical modelling, micromechanical structures are divided into discrete elements that are modelled using rigid-body dynamics. Some structural elements can be analytically modelled simply as rigid body masses, while other models have included the effects of bending, torsion, axial and shear stress. In the case of composite structures, an equivalent moment-of-inertia method has been applied to derive the compliance of the device in three orthogonal directions.

Before implementing numerical simulations, the heterogeneous section can be converted into a new shape with a homogeneous Young's modulus (E_n), using the equivalent-section method (Gere and Timoshenko, 1997). This is done, as an example, by normalizing each layer width to the Si_3N_4 (PAD layer) Young's modulus, which is the maximum value among the materials of the considered CMOS technology. The normalized cross-section, shown in Figure 5.1, corresponds to a homogeneous beam with the same mechanical properties as that of the initial cross-section.

The moment of inertia (I_n) of the normalized section is given as:

$$I_n = \sum \left[\frac{b_i t_i^3}{12} + S_i \left(h_n - h_i \right)^2 \right] \tag{5.3}$$

where b_i is:

$$b_i = b \frac{E_i}{E_n} \tag{5.4}$$

$S_i \left(= b_i t_i \right)$ represents the area of a given rectangular section in the normalized shape while h_n represents the vertical location of the neutral axis:

$$h_i = \sum_{k=1}^{i-1} t_k + \frac{t_i}{2} \quad , \quad i = 1, \ldots, N \tag{5.5}$$

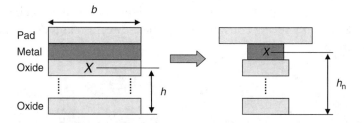

Figure 5.1   Cross-section normalization of a heterogeneous structure

$$h_{\mathrm{n}} = \frac{\sum_{i=1}^{N}(s_i - h_i)}{\sum_{i=1}^{N} s_i} \qquad (5.6)$$

As observed in the case of the simple, uniform and isotropic cantilever discussed in Chapter 1, given $l = L$, the generic linear dimension, M, scales as l^3 and K_z scales as l, or linearly with l. For example, if one scales the linear dimensions of the proof-mass size, spring length and width by a factor of 10, isomorphically, the corresponding scaled mass and elastic constant are approximately, respectively, $M' = 0.001\,M$ and $K'_z = 0.1\,K_z$. This means that the 10 times linearly scaled device is 1000 times lighter, but only 10 times less stiff than its non-scaled counterpart. A beneficial consequence of this scaling characteristic is hence that microaccelerometers can withstand highest accelerations without breaking or even being significantly disturbed. A negative consequence of the diminishing significance of inertial forces on the micrometre scale is that devices requiring proof-masses (e.g. accelerometers) must have motion-detection systems with a much higher sensitivity.

5.2.3 Accelerometers

The operation of a device as an inertial sensor will be considered in the following and the results obtained with prototypes realized through CMOS and SOI-based technologies will be compared. Such a system can be modelled from a mechanical point of view, as a spring-mass-damper, as represented in Figure 5.2.

If the system experiences an acceleration, the mass is subjected to a force proportional to the acceleration, which is contrasted by the spring elastic reaction and the damping affecting the dynamic of the movement, expressed as a function of the surrounding medium properties (viscosity, temperature). At equilibrium, the acceleration produces a net displacement of the proof-mass position.

The sensitivity of such an inertial sensor in terms of mass displacement versus acceleration can be again expressed as:

$$S = \frac{\mathrm{d}x}{\mathrm{d}a} = -\frac{M}{K} \qquad (5.7)$$

Then the spring constant scales linearly with the linear dimension; thus, if the simple inertial sensor described here is isomorphically scaled

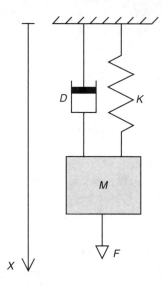

Figure 5.2 Modelling of an inertial sensor as a spring-mass-damper system

down by a factor of ten, its sensitivity to acceleration has a 100-fold improvement, i.e. the same acceleration produces a 100 times higher displacement or, with the same achievable displacement, one can measure accelerations 100 times smaller. Of course, the overall performance of an inertial sensor, as for any kind of sensor, will depend on the adopted readout strategy and circuits. However, this is no matter for this section and, moreover, it does not affect the generality of this discussion.

5.2.4 Resonant mass sensors

Resonant sensors have been well investigated in the past as devices for the measurement of several quantities, and they are attractive because of their potential stability and quasi-digital output (Hauptmann, 1991; Howe, 1987). In particular, resonant sensors can be used as highly sensitive mass sensors (Tilmans *et al.*, 1992; Meccea, 1994; Di Nucci *et al.*, 2003), which after surface functionalization with sorption coatings can be used for chemical sensing.

The realization of resonant microsensors in silicon requires the on-chip integration of actuating and sensing elements that provide vibration excitation and detection. For instance, some works present microresonators in silicon-on-insulator (SOI) technologies, where vibration actuation and sensing is accomplished electrically or electrostatically (Plaza *et al.*, 2002; Gigan *et al.*, 2002). The use of the piezoelectric principle for on-chip excitation and detection can be somewhat problematic due to process compatibility issues between silicon and piezoelectric materials.

Here a different approach is proposed, where microresonators are excited by an off-chip actuator made of a piezoelectric film of lead zirconate titanate (PZT) to which the resonator is bonded, resulting in a hybrid resonator structure. The PZT exciter element is realized as a screen-printed film in thick film technology (TFT) on alumina substrate (Ferrari *et al.*, 2001), that in this case acts as a wideband mechanical actuator with constant amplitude around the resonant frequency of the silicon-based structure. The proposed actuation strategy has an advantage in the simplicity of the sample preparation technique which does not require any additional layer for thermal/electrical actuators on the silicon microresonator. The resonator consists of a suspended element that changes the resonance frequency when its mass varies, therefore working as a 'microbalance'. The sensing of the vibrating element is performed on the chip by means of piezoresistors embedded into the resonator hinges. The design of the sensor and the actuator is independent and therefore a high resonant frequency and mass sensitivity can be obtained by optimizing the dimensions of the microresonator, without specific constraints on the off-chip excitation element. The actuator and the sensing devices can be connected in a closed loop configuration, realizing an oscillator that tracks the resonant frequency of the microresonator.

If a suitable excitation drives the device to vibrate at its own resonance frequency, ω, which is given by:

$$\omega = \sqrt{\frac{K}{M}} \tag{5.8}$$

it can be easily seen that $\omega = l^{-1}$, which means that for the linearly scaled device $\omega' = 10\omega$. The vibrating system may be used as a mass sensor by measuring its resonance frequency shift with respect to a reference, or 'unloaded', value ω_0, due to a change in the mass, especially if the variation of the spring mass can be neglected.

The mass sensitivity, expressed as the 'left-shift' of the resonance frequency to mass change is derived from the derivative of

ω with respect to m, as shown for a cantilever in equation (1.6) (Chapter 1):

$$S = \frac{d\omega}{dm} = -\frac{1}{2}\frac{\omega_0/M_0}{\sqrt{(1+m/M_0)^3}} \tag{5.9}$$

$S = l^{-4}$, which means that $S' = 10\,000\,S$, i.e. a linear scale factor of 10 in the dimensions of the square plate and beams leads to a 10 000-fold improvement in sensitivity of resonance frequency to mass change, which also means that smaller devices can potentially detect even smaller masses or mass changes.

Depending on the adopted technology, different dissipative contributions have to be included to model accurately device behaviour, thus always limiting performances. Two examples of devices realized in standard CMOS technology and SOI-based technology will discussed in the following to put into evidence scaling potentials and limitations of devices that can operate as accelerometers or mass sensors.

5.3 EXPERIMENTAL DEVICE PROTOTYPES

5.3.1 CMOS devices

Illustrations of five different prototypes after TMAH etching procedures are shown in Figure 5.3. Mechanical flexures compliant in the direction orthogonal to the plane of the device have been designed, while different design requirements must be taken into account in the case of laterally operating accelerometers, or gyroscopes where the springs have to be contemporarily compliant to both the drive and the sense direction (Zhang *et al.*, 1999; Clark *et al.*, 1996).

Since an optical readout strategy has been conceived for the proposed prototypes, holes typically necessary for improving TMAHW procedures action have been disposed at the edges of the membrane, thus leaving a 'transparent' central area where light can propagate to perform transmission measurements (Baglio *et al.*, 2001). Dimensions of the central plate are hence a compromise between the need to release completely the structure through TMAHW etching without placing holes in the central area (smaller structures can be easily released) and the need to have an active area where it is possible to perform optical measurements (not less than $100\,\mu m$ to use optical fibres in a laboratory measurement apparatus for devices characterization). Large structures,

Figure 5.3 Micrographs of CMOS membranes realized at DIEES, University of Catania. The proof-mass consists of transparent oxides since an optical readout strategy has been implemented for such sensors

overcoming the critical linear dimension of 500 μm for the proposed structures, will not be released.

It can be highlighted that limits to scaling are both imposed by technological aspects and the selected readout strategy.

5.3.1.1 Devices Fabrication Process

Furthermore, since a standard CMOS technology, described in Chapter 2, has been adopted for the fabrication of the prototypes shown in Figure 5.3, three different runs have been necessary to improve knowledge about the effects of micromachining on the process and materials, which remain neither characterized nor optimized for their electromechanical properties (Fedder, 1997). Scaling of the devices' geometrical features, aiming to optimize their behaviour both as accelerometers or resonant mass-sensors, cannot be, in fact, easily addressed without experiencing the unwanted effects of the actions of micromachining operations.

Moreover, the most common problem associated with fabricating membranes, beams or cantilever structures with metal or dielectric thin films is the magnitude and variance of the stress generated in the materials during fabrication. When the suspended plate is released, the internal stress causes it to bend in a concave or convex manner, depending upon the stress state (tensile or compressive) within the materials. This is, for example, a severe impediment to the advancement of micromechanical sensors with optical outputs, where surface flatness and uniformity are very important. In addition, surface flatness becomes more critical due to the decreased wavelength of the light beam for optical applications.

A particular low-pressure chemical vapour deposition (LPCVD) process which produces nearly stress-free, releasable, polysilicon structural thin films has been realized by the MUMPS fabrication service provided by the Microelectronic Center of North Carolina (MCNC) (Markus and Koester, 1996).

The high residual stress of metal layers, having a highest elastic constant, have been experienced with the devices of the first run, as shown in Figure 5.4. Trying to optimize the design of the thermal actuators, the metal wires, acting as collectors for the heat generated by the polysilicon heaters, were not anchored to the substrate, thus having one end anchored to the plate and the other 'free', even if sandwiched among the oxides. Probably during the last minutes of the TMAH etch, the metal curled significantly, together with the upper-lying VIA and

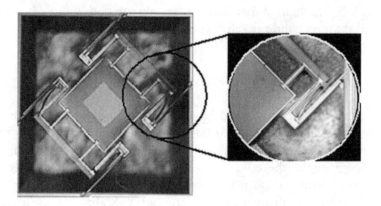

Figure 5.4 Micrograph showing the effects of the high residual stress of the metal layers, not achored to the substrate, after TMAH etching (2 h)

PAD layers, thus compromising the functionality of all of the devices, due both to the interruption of the polysilicon heaters and sometimes to the breaking of the freestanding flexures.

A different problem has been instead experienced from the device shown in Figure 5.5. Structural damage has been induced in the structure, probably due to the excessive length-to-width ratio of its sustaining springs which were not able to sustain the forces exerted during TMAHW etching.

Finally, even in the case of devices correctly released, the sustaining springs curl up out of the substrate plane, as shown in Figure 5.6,

Figure 5.5 Micrograph focusing on the damage induced during TMAHW etching procedures

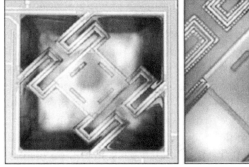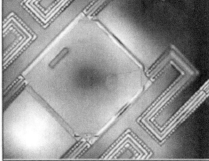

Figure 5.6 Micrograph of the device shown in Figure 5.3(b), of the 'second design'

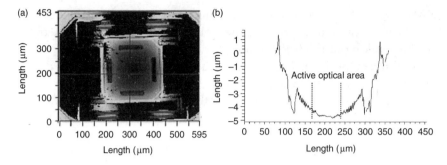

Figure 5.7 (a) Two-dimensional surface profile section of the device shown in Figure 5.3(b). Due to the residual stress during the etching procedure, the oxide membrane bends in a concave manner. (b) One-dimensional profile of the device shown in Figure 5.3(b), after releasing

due to internal stresses in their thin structural layers. As expected, the residual stress during the TMAHW etching induced a deformation of the suspended plate, as shown in Figure 5.7. The etched plate is hence slightly elevated above the substrate plane at its corners, where a metal ring has been realized. The flatness of the exposed surface has been measured by using a Wyko interferometric profiler microscope. The peak-to-valley of the unactuated plate is about $5\,\mu m$, but it must be considered as an active region of $50\,\mu m$ in the central area of the plate where the peak-to-valley length is $0.63\,\mu m$.

5.3.2 SOI devices

5.3.2.1 Fabrication Process and Experimental Prototypes

The SOI microresonator has been realized with a custom process based on the SOI wafer and a few other materials needed for piezoresistor realization. The fabrication process requires, in the simplest case where only the silicon wafer is processed, only three photolithographic masks, two for the mechanical structure definition and one for the encapsulating cap glass on the bottom of the die. Two further process steps have to be considered for the device which includes polysilicon and metal functional layers. The technology used is based on an N-type (100) oriented, $450\,\mu m$ thick BESOI (bond and etch back silicon on insulator) wafer. The thickness of the buried silicon oxide and of the upper c-Si layer is $2\,\mu m$ as shown in Figure 5.8. Several slightly different devices

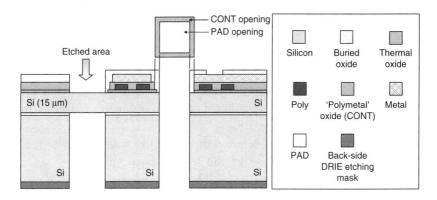

Figure 5.8 Front-side and back-side bulk micromachining in SOI technology

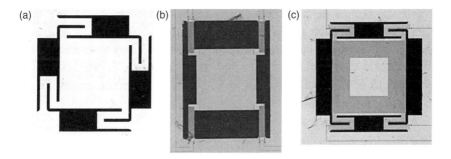

Figure 5.9 Layouts of three SOI-based sensors with different flexures and materials embedded into the suspended plate: (a) no functional materials are included; (b) polysilicon piezoresistors are embedded in the arms; (c) a suitable polysilicon heater is selectively embedded into the proof-mass (Baglio et al., 2005) © 1996 IEEE. Reproduced by kind permission of IEEE

have been realized. In Figure 5.9, three different layouts of the realized structures are reported (Baglio *et al.*, 2005).

The effect of the residual stress, during releasing of the structures and the spreading of the mechanical parameters, as the elastic constants, has been evaluated for both of the two different SOI wafers, having a 5 μm and 15 μm upper cSi layer thickness, respectively. To this purpose, an increased residual stress, in the case of the wafers with a upper c-Si layer 5 μm thick, was expected and has been confirmed by the first observations.

Both front-side and back-side DRIE etching procedures, 'stopped' by the 2 μm thick buried oxide layer, have been performed to release the structures. Moreover, the adopted etching strategy has allowed us to realize very flat surfaces. Therefore, two substantial differences arise between the devices realized by using the considered technologies, consisting of:

(a) The order of magnitude of the proof-mass, considerably higher in the case of the SOI process, where, due to a custom back-side and front-side DRIE procedure, the silicon wafer can be a part of the plate, as shown in Figure 5.8.
(b) When performing optical sensing, the operating mode, being a 'transmissive mode' in the case of the CMOS-based prototypes, using transparent materials in the visible wavelength range, is now a 'reflective mode' due to the opacity of the silicon substrates (both the c-Si upper layer and the wafer).

For both classes of structures, four identical piezoresistors were also included, one in each suspending folded beam to detect the out-of-plane displacement of the central mass. The microplate moves upwards and downwards like a piston when 'solicited' at its own first resonance frequency. More details on this behaviour can be predicted by *Finite Element Analysis*.

A thermally selective deposition strategy has been developed in order to avoid further masking for the deposition of the active material in the central area of the plate. A polysilicon heater is also included in order to possibly 'warm-up' the structure during the deposition of a suitable polymer, as shown in Figure 5.9(c). Some other strategies can be thought to functionalize the mass surface in order to absorb different compounds and to perform chemical sensing using selectively coated surfaces. Such a device can be hence used as a mass sensor in applications where the mass variation due to the sorption of a particular compound is used to determine physical or chemical properties of that material.

Finally, a glass cap processed through DRIE etching have been bonded to the bottom of the die, with openings in correspondence of the devices active operating area, prevalently to protect the die. At the same time, a large gap between the proof-mass and the same glass cap, being approximately $80\,\mu m$, exists, allowing us to neglect dissipative effects such as 'squeeze-film' damping. Obviously for devices in which back-side DRIE removes the $450\,\mu m$ substrate under the proof-mass, such a gap is approximately $530\,\mu m$, meaning that the hypothesis to neglect 'squeeze-film' damping has a higher validity.

A hybrid system, composed of a micromachined SOI-based element bonded to a PZT screen-printed exciter on an alumina substrate, has

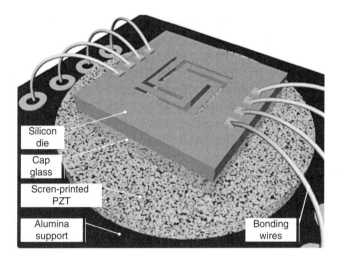

Figure 5.10 Schematic view of the hybrid system (Baglio *et al.*, 2005) © 1996 IEEE. Reproduced by kind permission of IEEE

Figure 5.11 Experimental prototype after wire-bonding of the SOI-MEMS onto the PZT exciter (Baglio *et al.*, 2005) © 1996 IEEE. Reproduced by kind permission of IEEE

been conceived to provide energy to the microresonator operated as a mass sensor, and is schematically shown in Figure 5.10.

The device is again conceived to have parallel actuation and sensing axes along the direction normal to the device plane – an experimental prototype is shown in Figure 5.11. The actuation energy, often being a significant issue related to the realization of similar devices, is provided by the piezoelectric electromechanical transducer, uncoupled from the sensing element. The mass variation of the microresonator, due to the chemical or physical compounds sorption on its surface, will induce variations in the resonance frequency of the mass sensor, without affecting the exciter dynamic behaviour.

5.3.3 Finite element modelling

Although finite element analysis is not yet a suitable designing instrument for engineers, nevertheless it is actually largely used for MEMS development since it allows us to obtain results with a degree of accuracy in the order of the approximation of the numerical methods to evaluate electrical and mechanical nonlinearity.

It has been used here to determine the mechanical modes of the structures due to external forces. Simulations have been performed using the commercial software 'ANSYS 6.0', in order to compare results with those of the analytical method. We used a linear finite element analysis with a discretization made by about 100 000 tetrahedral elements of the tenth order in the case of CMOS devices. In fact, the composite structure obtained in this case, with very thin layers, does not allow us to 'mesh' the device with less elements without violating rules about the form factors. A different approach is followed for the realized SOI devices, where layers thicknesses allow 'meshing' with typically 20 000 tetrahedral elements.

Animated depictions, reported in Figure 5.12, of the first five resonance frequency modes, for device 1 (see Figure 5.3(a)) in CMOS technology, reveal the differences of motions for the different modes

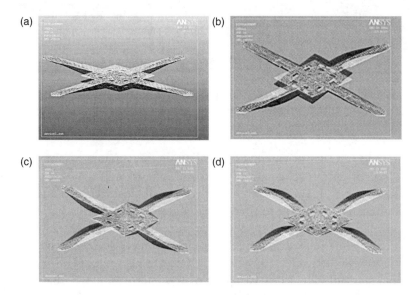

Figure 5.12 Deformed shapes for device 1 (see Figure 5.3(a)) at different excitation frequencies (Baglio *et al.*, 2005) © 1996 IEEE. Reproduced by kind permission of IEEE

excited. Regarding device 1, for the first mode's motion, we can see the entire microplate moving upwards and downwards like a piston, at a frequency of 29.1 kHz. The sustaining springs seem to bend correctly and the conceived behaviour of the device is obtained at a relatively low frequency. From the theoretical calculations, a resonance frequency of about 25.7 kHz has been evaluated. The discrepancy in the results can be justified by the different approach to the problem, being distributed in the case of finite element modelling, while a 'lumped' parameters model has been considered for theoretical calculations. The second and the third mode (the latter not shown) are symmetrical rotations around the torsion x-axis and y-axis, respectively. A couple of opposite springs bend, while the others are rotating. They, occur, respectively, at 82.4 kHz and 82.5 kHz. The fourth mode's motion is interesting only the freestanding beams, thus not soliciting the plate to move; the opposite springs of the microplate synchronously flap up-and-down, similar to a bird in flight, and occur at 136 kHz. The fifth mode's motion, occurring at 184 kHz, is similar to the previous one but in this case all of the springs flap synchronously, while the plate is not moving appreciably.

Regarding device 2 (see Figure 5.3(b)), the first mode's motion, shown in Figure 5.13(a), is again the conceived one and has the entire microplate moving upwards and downwards like a piston, at a frequency of 46.1 kHz. From the theoretical calculations, a resonance frequency of about 41.2 kHz has been calculated in this case. All of the other four modes' motions occur, respectively, at 65.4, 79.8, 96.9 and 100.9 kHz,, with the springs moving in different ways, causing in the first two cases rotation of the plate to the x-axis and the y-axis, respectively, while the displacement of the plate is not appreciable in the other two cases for the last two modes' motion.

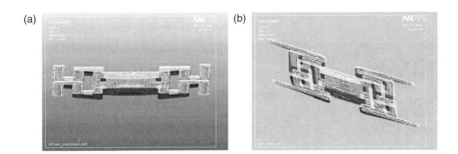

Figure 5.13 Deformed shapes for (a) device 2 (see Figure 5.3(b)) and (b) device 3 (see Figure 5.3(c)) at 46.1 and 24.0 KHz, respectively

Device 3 (see Figure 5.3(c)), shown in Figure 5.13(b) exhibits a resonance at a frequency of 24 kHz. From the theoretical calculations, a resonance frequency of about 21.2 kHz has been calculated, in this case by the analytical method. For all of the other four modes' motions, torsion is experienced in the second mode (65.4 kHz) and the third mode (79.8 kHz); it bends in a concave manner when excited at the frequency of the fourth mode (96.9 kHz), while the displacement of the plate is not appreciable in the case of the fifth mode's motion (100.9 kHz).

At the frequencies exciting the first mode for all of the considered devices, the maximum displacement of the microplate occurs in the z-direction. The respective values are reported in Table 5.1.

Similar simulations have also been performed for SOI devices, as shown in Figure 5.14. This type of device is obviously at its own resonance frequency in which a piston-like motion is experienced. Torsions and 'spring flappings' must be carefully avoided.

Some conclusive remarks can be made about the role of finite element analysis (FEA) for designing these kind of micromechanical devices. Although accurate, these simulations are slow and computationally intensive, both in memory and time. Thus, while these approaches are suitable for accurately simulating simple structures, they are not powerful instrument for simulating large three-dimensional structures with multiple segments. Considerable experience is required on the part of the user for the solid modelling and 'meshing' operations, thus not

Table 5.1 Maximum displacements occurring for the considered devices in the z-direction

Device	Maximum displacement (nm)
1	42
2	147
3	220

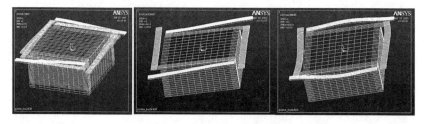

Figure 5.14 Deformed shapes of a reference SOI sensor at different excitation frequencies

allowing analysis of all of the possible alternatives that a designer can consider during these projects. Anyway, FEA is really a powerful instrument when verification of some models has to be addressed, so allowing us to better understand the behaviour of the examined devices, or when some boundary conditions or constitutive parameters change slightly, meaning that limited parametric analysis can be performed.

5.4 SCALING ISSUES ON MICROACCELEROMETERS AND MASS SENSORS

Considering a reference device, such as that schematized in Figure 5.15, for both the discussed microtechnologies, allows us to appreciate how the scaling devices' dimensions suggest their use for a particular application, rather than others. Figure 5.16 shows the scaling of a CMOS prototype with the following geometrical features: the proof-mass is a square with side length l, as well as the spring length; the spring width is assumed to be $l/6$, while the thickness is t.

If l varies from 600 to $200\,\mu m$, where limits are fixed by the micromachining procedures, the mass sensitivity increase significantly, going from approximately 500 to $24\,kHz/\mu g$. Therefore, the device is an optimum candidate to be operated as a very high sensitive mass sensor, over all of the considered dimension range. Its nominal mass sensitivity is, in fact, more than one order of magnitude higher than that of reference devices in such fields, being a quartz crystal microbalance (QCM) and resonant piezolayer (RPL) (Ferrari *et al.*, 1997) sensors that can

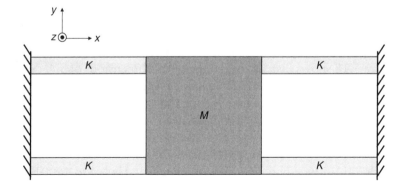

Figure 5.15 Simplified schematic of a reference mechanical sensor

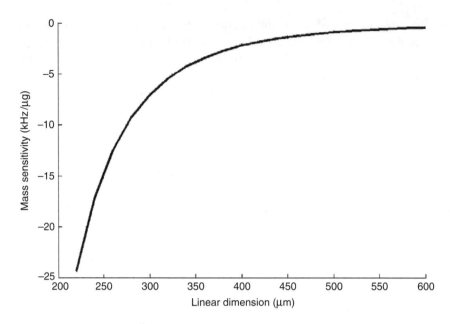

Figure 5.16 Mass sensitivity of a CMOS reference device, scaling the linear dimension, l, from 200 to 600 μm

achieve such values at much higher frequencies, of the order of several megahertzs.

On the other hand, even if it is confirmed that sensitivity to static acceleration decreases with the shrinking dimensions in the same range, such a class of device will not give satisfying results if operated as an accelerometer, as shown in Figure 5.17 .

In both cases, the equivalent section moment has been implemented to accurately calculate the elastic constant of the springs, thus considering the composite nature of the structure, whereas dissipative contributions, as squeeze film damping (Starr, 1990), have been neglected due to the large gap under the plate, being approximately 200 μm.

Figure 5.18 shows instead the scaling linear dimensions of SOI devices, corresponding to that depicted in Figure 5.16, which suggests their use as microaccelerometers. Their geometrical features are as follows: proof-mass is a square with side length l, as well as the spring length, while the spring width is assumed to be $l/6$, with thickness t.

If l varies from 1500 μm to 450 μm, the mass sensitivity increases again significantly, going from approximately 50 Hz/μg to 4500 Hz/μg, in the case shown in Figure 5.19. The minimum side length is fixed at a value corresponding to the proof-mass thickness, varying approximately

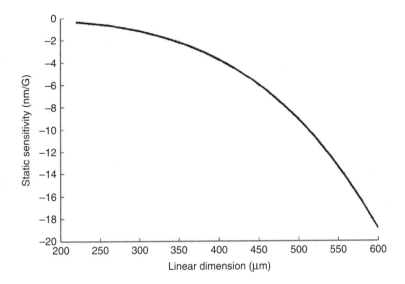

Figure 5.17   Static sensitivity to acceleration of a CMOS reference device, with the scaling linear dimension, l, ranging from 200 to 600 μm

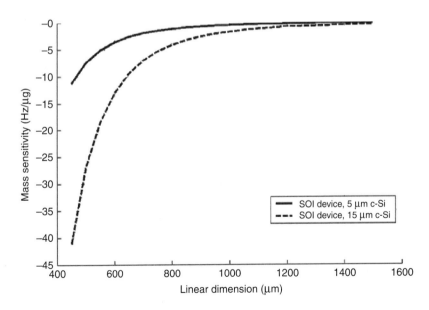

Figure 5.18   Mass-sensitivity of SOI reference devices, where the 450 μm-thick substrate *has not been* 'under-etched' from the proof-mass; scaling linear dimension, l, from 450 to 1500 μm

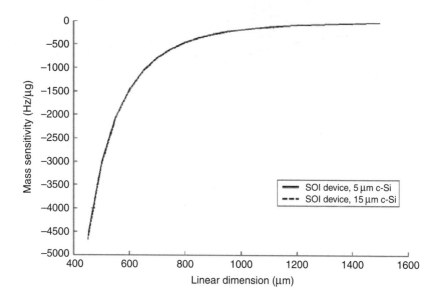

Figure 5.19 Mass sensitivity of SOI reference devices, where the 450 μm-thick substrate *has been* 'under-etched' from the proof-mass; scaling linear dimension, *l*, from 450 to 1500 μm

from 458 μm to 468 μm. Otherwise, the thickness overcomes the length and width of the proof-mass and torsional modes may occur very easily. Moreover, the CMOS devices have better performances if operated as mass sensors, even if adequate SOI devices show a very high sensitivity towards the minimum linear dimensions, if compared again with the QCM system.

In particular, in Figure 5.18 a comparison between the results obtained from equal SOI devices realized through two different wafers, in which only the thickness of the upper c-Si layer is 5 μm or 15 μm, is reported. A silicon substrate is included in the proof-mass.

The masses scale approximately in the same way, due to the prevalent role of the silicon substrate on the proof-mass value, while the elastic constant of the springs is always larger in the case of a wafer with an upper c-Si layer which is 15 μm thick.

The mass sensitivity increases drastically if such silicon substrates are removed from the proof-mass. In addition, they are now comparable for both of the considered wafers, over the considered scaled dimensions range, as shown in Figure 5.19.

In contrast, SOI devices with the silicon substrate included in the proof-mass have potential to be very highly sensitive accelerometers.

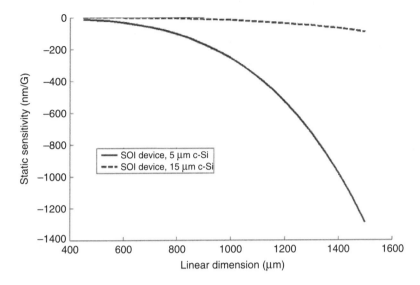

Figure 5.20 Static sensitivity to acceleration of SOI reference devices, where the 450 μm-thick substrate *has not been* 'under-etched' from the proof-mass; scaling linear dimension, *l*, from 450 to 1500 μm

Shrinking dimensions deteriorate the performances in such cases; moreover, suitable geometrical features lead to sensitivity to static acceleration of the order of 1300 nm/G. The results reported in Figures 5.20 and 5.21 suggest that better performances are obtained with devices in which the c-Si layer is 5 μm thick. The elastic constant of the spring, being at the denominator of the static sensitivity for accelerometers, is always lower than in the case of the wafer with a 15 μm thick c-Si layer.

5.5 SOME EXPERIMENTAL RESULTS

Preliminary characterization of the realized SOI prototypes has been performed by processing the electrical signals coming from the piezoresistors embedded within the resonator arms (Savalli *et al.*, 2004). The conditioning circuit shown in Figure 5.22 has been realized, where two of the four piezoresistors have been used in a Wheatstone bridge connected to an instrumentation amplifier. The microresonator has been characterized versus frequency by mounting the device on a calibration exciter having a built-in reference accelerometer. The frequency response and the conditioning circuit output have been

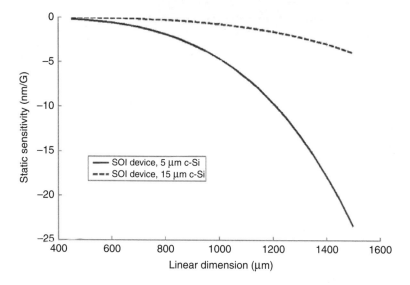

Figure 5.21 Static sensitivity to acceleration of SOI reference devices, where the 450 μm-thick substrate *has been* 'under-etched' from the proof-mass; scaling linear dimension, *l*, from 450 to 1500 μm

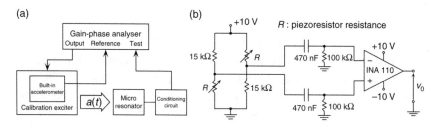

Figure 5.22 (a) Block diagram of the measurement set-up used for SOI devices characterization as mass sensors. (b) The corresponding conditioning circuit (Baglio *et al.*, 2005) © 1996 IEEE. Reproduced by kind permission of IEEE

measured with a gain-phase analyser. The block diagram of the measurement set-up is also shown.

Figure 5.23 shows the frequency response measured over the range 1–4 kHz, with resonance peaks corresponding to different vibration modes. A quality factor Q of about 440 has been measured, which confirms the expectations for a high value, considering the high compliance of the microstructure and a low value for the damping factor, due to the large value of air gap under the proof-mass for such classes of devices. Instead of performing measurements of frequency shift versus mass increase, a different approach has been initially followed to test the microstructure. An SOI microresonator bonded to the PZT exciter was

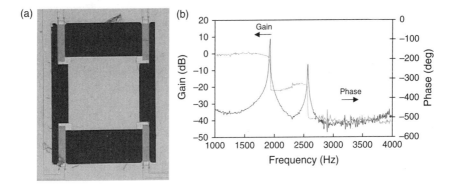

Figure 5.23 (a) Experimental SOI prototype and (b) measured frequency response (Baglio *et al.*, 2005) © 1996 IEEE. Reproduced by kind permission of IEEE

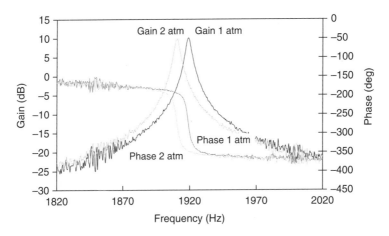

Figure 5.24 Experimental resonance frequency down-shift due to the increasing hydrostatic pressure (Baglio *et al.*, 2005) © 1996 IEEE. Reproduced by kind permission of IEEE

enclosed within in a sealed package and the pressure was changed from the atmospheric value up to 2 atm. The increase in the hydrostatic pressure, as expected, determines a rise in the equivalent vibrating mass of the microresonator and therefore a decrease in the resonant frequency, as shown in Figure 5.24.

In addition, an increase in the damping can also occur due to viscous friction. The expected frequency downshift was experimentally confirmed, as shown in the results of Figure 5.24, while the increase in damping appears to be negligible.

In order to realize a compact measuring system, in which the proposed mass microsensor is coupled with a suitable conditioning circuit, a

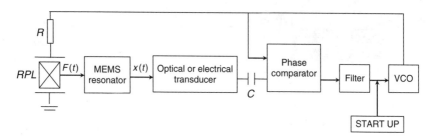

Figure 5.25 Block diagram of the PLL-based frequency-tracking oscillator (Baglio *et al.*, 2005) © 1996 IEEE. Reproduced by kind permission of IEEE

closed-loop configuration has been designed to measure the variation of frequency as a function of mass variation. In this way, high benefit is obtained for the frequency-conditioning circuits in the open-loop chain, as PLL, in tracking the matching conditions between the sensor output and the voltage controlled oscillator (VCO) output, as shown in Figure 5.25. The VCO feeds the voltage divider formed by R and the *RPL* impedance with a square-wave whose frequency is controlled by the averaged output voltage of the phase comparator. That latter block is an 'edge-triggered' network which senses the phase difference between the VCO and MEMS output and forces it constantly to zero, due to feedback action.

The circuit 'locks' at the frequency where the microsensor impedance is purely resistive (AG 1), referring to the impedance of the equivalent electrical model, and higher in value, i.e. at around $f(AG\ 2)_n$, and automatically tracks the resonance condition during its variations due to the measurand action.

Concluding, several advantages come from the proposed system as uncoupling actuation and sensing, as well as to realize low-cost devices which could be arranged in a matrix structure where each pixel oscillates at its own resonance frequency, thus realizing 'a kind' of mass spectrometer.

5.6 VIBRATING MICROGYROSCOPES

A variety of sensing techniques, including optical, capacitive, tunnelling and piezoresistive have been used to estimate the Coriolis force, and hence the rotation rate, by measuring the displacement of the proof-mass in a direction orthogonal to both the driven motion and the axis about which rotational motion is to be sensed (Yazdi *et al.*, 1998; Hashimoto *et al.*, 1995; Maenaka and Shiozawa, 1994; Soderkvist,

1996; Clark *et al.*, 1996; Oh *et al.*, 1997; Tanaka *et al.*, 1995; Ayazi and Najafi, 1998).

Several different technologies, etching procedures and device configurations have been proposed for realizing resonant vibrating micromachined gyroscopes, highlighting the common critical issues of detecting displacements along the sense axis that are orders of magnitude less than the drive amplitudes and for keeping the drive and sense modes uncoupled (Xie *et al.*, 2002; Xie and Fedder, 2003; Acar and Shkel, 2001).

The design of such a device must hence focus on obtaining nearly equal compliance of the proof-mass along two orthogonal directions.

If a lateral-axis vibrating configuration is considered to sense the angular rate of a rotating object, a massspringdamper equivalent system, where the Coriolis force is assumed as the inertial force transferring the energy from the driving mode to the sense mode, can be used for modelling the system, as illustrated in Figure 5.26, while a photomicrograph of a vibrating gyroscope realized through a standard 'CMOS AMS' $0.8\,\mu\text{m}$ process is shown in Figure 5.27.

The equations of motion can be expressed in the following forms:

$$F_{\text{drive}} = M\ddot{x} + D_x\dot{x} + k_x x$$
$$F_{\text{Coriolis}} = M\ddot{y} + D_y\dot{y} + k_y y \tag{5.10}$$
$$F_{\text{Coriolis}} = \left|2m\vec{\Omega}_z(t) \times \dot{x}(t)\right| = 2\,m\Omega_z(t)\omega_x A\cos\,(\omega_x t)$$

where Ω is the angular velocity and $x(t) = A\sin\,(\omega t)$.

The Coriolis force is hence an amplitude-modulated signal where the carrier frequency is the oscillation frequency and the rotation rate modulates the amplitude. Then, a dual sideband signal centred on the

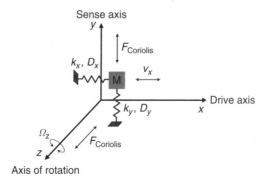

Figure 5.26 Mechanical model for linear vibrating gyroscopes

Figure 5.27 Micrograph of a vibrating gyroscope realized through a standard 'CMOS AMS' 0.8 μm process

oscillation frequency is observed for such a force. Since the y-axis force is proportional to velocity, the motion of the proof-mass is elliptical.

Moreover, the general solution for $y(t)$ can be found and it describes the displacement along the sense axis of the gyroscope (Baglio *et al.*, 2004b). The main, well-known, result is that for driving frequencies less that w_r the sense mode oscillates approximately in phase with the drive frequency and for driving frequencies greater than w_r the phase lag is π.

Therefore, in order to define gyroscope performances two fundamental parameters, such as sensitivity and resolution, can be analysed. In addition, the sensor output is required to be insensitive to parameters other than the measurand, especially environmental parameters such as temperature and pressure. If an open-loop displacement sensing strategy is considered, the sensitivity can be expressed as a ratio of the measured displacement $y(t)$ in the sensing direction to the rotation rate (Ω_z) to be measured (Seshia *et al.*, 2002):

$$S = \frac{y}{\Omega_z} = \frac{2A\omega_x}{(\omega_y^2 - \omega_x^2) + j\left(\frac{\omega_x\omega_y}{Q_y}\right)} \quad (5.11)$$

Gyroscopes that utilize displacement sensing to determine the rotation rate often operate under the conditions of matched modes or closely matched modes for improved resolution and sensitivity under low-pressure (high-quality factor) conditions. If such matching conditions are obtained, a gain of Q in the system response can be observed. However, perfect mode matching typically limits the bandwidth of the input rate to less than a few hertz which is unsuitable for many applications. On the other hand, an open-loop implementation with slightly mismatched modes trades-off sensitivity for bandwidth. In addition, since the sensitivity is still inversely proportional to the difference between the drive and sense mode natural frequencies, and these frequencies may be different functions of the influencing quantities, such as temperature and pressure, potential bias instability and sensitivity drift can be observed (Seshia *et al.*, 2002).

Provided that there is a means to tune the y-axis resonant frequency, it is desirable to operate the gyroscope with a 5–10 % frequency mismatch,, yielding a gain of 5–10 in sensitivity. Reducing the mismatch would increase the sensitivity but is probably difficult to maintain for any length of time (Clark *et al.*, 1996).

At the limit in which the difference approaches zero, the vibratory motion can also be distorted by mechanical coupling between the two modes (Kawai *et al.*, 2001). This distortion increases the 'noise' in the sensor output and degrades the resolution of the angular rate. Therefore, in order to improve the resolution when the sensitivity is enhanced by reducing the frequency difference between the driving and sensing modes, it is necessary to reduce the mechanical coupling.

On the other hand, mechanical coupling deriving from unavoidable imperfections in the microstructures' realization process implies that the proof-mass oscillates along an axis that is not exactly parallel to the drive axis. Then, asymmetry of the microstructures, unbalance of stress on the beams of the microstructure and unbalance of electrostatic forces for driving the device are the most important causes of mechanical coupling. A small fraction, ε, of the oscillation can hence be supposed to lie along the sense axis. Considering that such a portion of the displacement along the sense axis can be expressed as (Clark *et al.*, 1996):

$$y(t) = \varepsilon x(t) \tag{5.12}$$

then this acceleration along the sense axis can be defined as:

$$\ddot{y}_{mc}(t) = -\varepsilon x(t) = \varepsilon A \omega_x^2 \sin\left(\omega_x t\right) \tag{5.13}$$

This acceleration term is referred to as 'quadrature error' in Clark *et al.*, (1996). It is very similar to the Coriolis acceleration, even if they can be distinguished by their phase relative to the driven oscillation. It can be very large and looking at the ratio between the Coriolis acceleration and that coming from the 'of-axis' motion, some considerations can be drawn:

$$\frac{\ddot{y}_{Coriolis}}{\ddot{y}_{mc}} = \frac{2\Omega_z \omega_x A \cos (\omega_x t)}{\varepsilon A \omega_x^2 \sin (\omega_x t)} \approx \frac{2\Omega_z}{\varepsilon \omega_x} \qquad (5.14)$$

Considering a rotation rate of $\Omega = 1$ deg/s and an oscillation frequency of 30 kHz, the two accelerations are comparable (ratio ≈ 1) for $\varepsilon \approx 1.8 \times 10^{-7}$, hence meaning that the oscillation direction must be imposed very precisely.

Closed-loop control can be a solution to these issues by extending the bandwidth without significantly degrading the resolution. However, any control implementation is compounded by the severe challenge of resolving 'sub-angstrom' displacements and maintaining system stability in the presence of larger perturbations (such as coupling of the driven motion of the proof-mass) for a single or multi-degree of freedom system (Shkel *et al.*, 1999).

5.6.1   Coupled vibratory gyroscopes

Apart from discussions regarding noise sources for gyroscopes, which can be found in (Clark *et al.*, 1996), a novel approach based on arrays of coupled resonant vibrating gyroscopes will be proposed in the following. It is clear that unavoidable imperfections in realizing the microstructures and slight mismatches in precisely setting the driving frequency typically affect these systems. Such variations can be efficiently neglected by synchronizing an array of ideally identical gyroscopes, both in phase and frequency (Mochida *et al.*, 2000; Alper and Akin, 2002).

A linear chain with periodic boundary conditions, is shown in Figure 5.28, where each gyroscope connected to its two nearest neighbours can be considered.

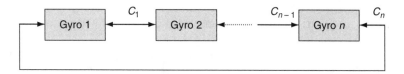

Figure 5.28   Illustration of a chain of gyroscopes with bidirectional coupling (Baglio *et al.*, 2004b). Reproduced by kind permission of EUROSENSORS 2004

The design of an array of non-identical gyroscopes usually involves choosing a quality factor for the sense axis that minimizes phase mismatch in the array and maximizes the amplitude response to the Coriolis driving force.

The goal here is to use a coupling network to minimize the phase offset so that the output from each gyroscope can be summed together. This make the output from the non-identical array similar to the identical array. There are other alternative solutions (e.g. using a phase-shifting system) but the coupling network works pretty well and it might be simpler to implement. Such a strategy allows us to 'force' phase-synchronization.

Regarding the possibility of using demodulators for the open-loop sensing strategy, a noticeable advantage is obtained if all of the gyroscopes synchronize. Capacitive sensing can be performed, thus linearly summing all of the sinusoidal outputs of the gyroscopes (linear superposition).

For three identical gyroscopes with identical couplings (C), the model represented by equations 5.15 can be considered:

$$m\ddot{x} = -k_x x - d_x \dot{x} + F_{\text{drive}} \tag{5.15a}$$

$$m\ddot{y} = -k_y y - d_y \dot{y} + F_{\text{Coriolis}} + F_{\text{coupling}} \tag{5.15b}$$

$$F_{\text{Coriolis}} = |2m\Omega_z \times \dot{x}| = 2m\Omega_z A_d \omega_d \sin(\omega_d t), \quad \text{if} \quad \dot{x} = A_d \cos(\omega_d t) \tag{5.15c}$$

$$F_{\text{coupling}} = C(y_{\text{neighbour}} - y_{\text{self}}) = C(y_{i+1} - y_i) + C(y_{i-1} - y_i) \tag{5.15d}$$

An amplitude-modulated sinusoidal driving force has been taken into account for this system. The coupling term is introduced in to the model, which has the form $C(y_{\text{neighbour}} - y_{\text{self}})$, where y is the sense-mode displacement and C is the coupling strength. A coupling coefficient of zero corresponds to an uncoupled system.

Damping and restoring coefficients have been evaluated for each axis by means of analytical models and compared with results obtained from finite element analysis. The nominal values for the mass, m, linear restoring force, k, damping, D, drive resonance, ω_r, external angular rate, Ω, drive amplitude, A_d, and drive frequency, ω_d, are reported in Table 5.2.

A ring configuration composed by three elements has been considered as the basic configuration for all of the simulations. Then, in the matrix form, masses, damping and restoring coefficients can be expressed by matrix (3) if the coupling term is included in the model:

Table 5.2 Simulation parameters (Baglio *et al.*, 2004b). Reproduced by kind permission of EUROSENSORS 2004

Parameter	Value	Unit
m	10^{-9}	kg
k	2.64	N m^{-1}
D	5.14×10^{-7}	N s m^{-1}
w_r	51 472	rad s^{-1}
Ω	0.15	rad s^{-1}
A_d	10^{-6}	m
C	0–1	—

$$M = \begin{bmatrix} m_1 & 0 & 0 \\ 0 & m_2 & 0 \\ 0 & 0 & m_3 \end{bmatrix} \quad D = \begin{bmatrix} d_1 & 0 & 0 \\ 0 & d_2 & 0 \\ 0 & 0 & d_3 \end{bmatrix} \quad K = \begin{bmatrix} k_1 - 2C & C & C \\ C & k_2 - 2C & C \\ C & C & k_3 - 2C \end{bmatrix}$$
$$(5.16)$$

The effect of coupling on phase difference has been first examined by 'forcing' the system with an harmonic input, thus looking at its steady-state response. The results obtained from the simulations are reported in Figures 5.29–5.31.

Therefore, it is possible to see that the coupling network is able to decrease the sum of the phase differences from a maximum of 313.31 to

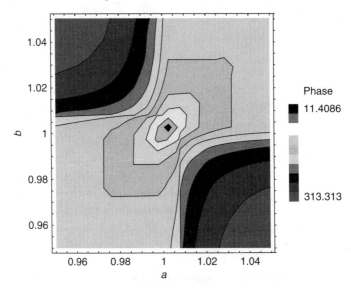

Figure 5.29 Effect of coupling on phase difference for three coupled gyroscopes ($C = 0$) (Baglio *et al.*, 2004b). Reproduced by kind permission of EUROSENSORS 2004

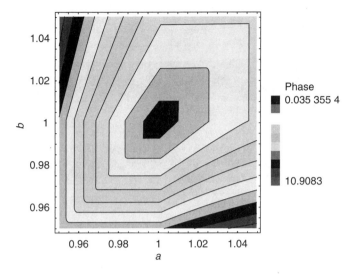

Figure 5.30 Effect of coupling on phase difference for three coupled gyroscopes ($C = 0.1$) (Baglio *et al.*, 2004b). Reproduced by kind permission of EUROSENSORS 2004

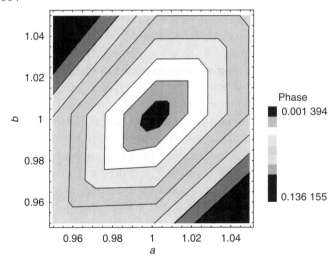

Figure 5.31 Effect of coupling on phase difference for three coupled gyroscopes ($C = 1$) (Baglio *et al.*, 2004b). Reproduced by kind permission of EUROSENSORS 2004

0.13 degrees. The masses has been varied by up to $\pm 5\%$ with respect to the nominal values assumed in Table 5.2. In particular, $m_1 = m$, $m_2 = a \times m$ and $m_3 = b \times m$ are assumed.

Increasing of the coupling strength does not significantly change the amplitude response, which is primarily determined by mass distribution.

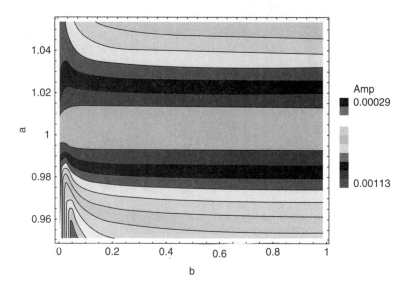

Figure 5.32 Effect of coupling on amplitude for three coupled gyroscopes, with varying couplings and masses (Baglio *et al.*, 2004b). Reproduced by kind permission of EUROSENSORS 2004

The effect of coupling on the amplitude for three coupled gyroscopes, where $m_1 = m$, $m_2 = m$ and $m_3 = a \times m$, is reported in Figure 5.32.

It can be highlighted that an improvement in the amplitude response and synchronization time for the array of gyroscopes could be achieved by means of the coupling term when spreading on initial conditions, gyroscope masses and driving frequencies are considered at the same time.

A random set of initial conditions have been imposed to the system. In particular a random 5 % variation for mass, and a random distribution between from 0 up to 1.5×10^{-9} m, and 1×10^{-5} m/s, with equal probabilities, for positions and velocities, respectively, have been set (Fogliatti, 2002). These random initial conditions place the gyroscope ring in a very disordered state, as shown in Figure 5.33.

Simulations, performed by using Matlab and Simulink, started by comparing the ideal case, in which all of the gyroscope are identical, with a configuration in which a random variation of the gyroscopes masses is considered. A constant external angular rate has been assumed in this phase.

In Figure 5.34, the summed temporal gyroscopes responses in the case of identical and non-identical masses (5 % spreading) with *no* coupling are reported.

For the 'ideal' system, no improvement in array synchronization time or amplitude response is realized by using coupling. Coupling

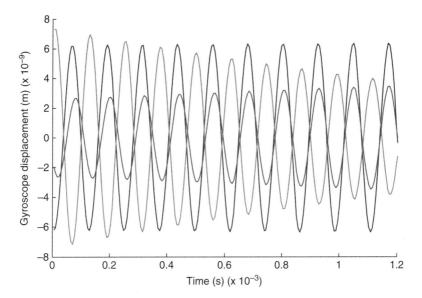

Figure 5.33 Time series for non-identical gyroscopes where no coupling is applied (Baglio *et al.*, 2004b). Reproduced by kind permission of EUROSENSORS 2004

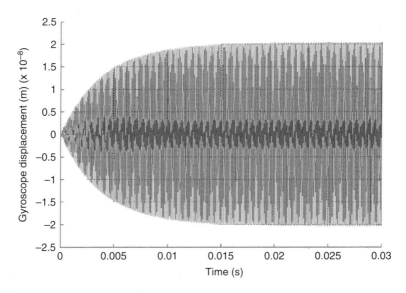

Figure 5.34 Temporal summed outputs of the gyroscopes – the summed output is evidently smaller for a system with three different masses and no coupling. These are driven at a resonance frequency of 51 472 rad/s (Baglio *et al.*, 2004b). Reproduced by kind permission of EUROSENSORS 2004

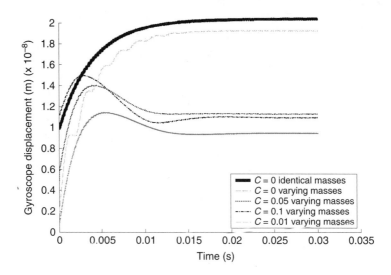

Figure 5.35   Effect of coupling on the demodulated responses from non-identical gyroscopes (Baglio *et al.*, 2004b). Reproduced by kind permission of EUROSENSORS 2004

decreases the amplitude response for an array of gyroscopes with identical resonant and drive frequencies. The summed and demodulated outputs are instead plotted in Figure 5.35, in the case of a non-identical gyroscope array.

Synchronization is lost in this latter case. Each sense-axis mode has a slightly different resonance frequency due to a variation in the mass of each gyroscope. The effect of coupling changes the response of the array which synchronizes, even if it does decrease the maximum summed demodulated amplitude. A maximum value can be observed in the case of identical gyroscopes, whereas it strongly decreases for a system with three different masses and $C = 0$. The maximum amplitude response of the coupled system is greater than the uncoupled system.

In Figure 5.36, the effect of a random variation of about 2.5 % in the mass is examined. This mass variation changes the resonant frequency of the sense mode to $8192 \pm 79\,Hz$. This system now does not have a single resonance frequency to exploit. Two values, slightly above and below the nominal resonance frequency, have been chosen for the drive frequency in order to examine the response of the array.

In Figure 5.37, the effect of the driving frequencies spreading and coupling strength tuning on the amplitude responses can be observed for an array of non-identical gyroscopes.

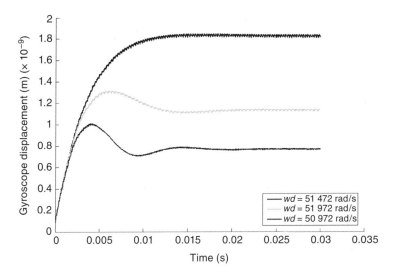

Figure 5.36 Non-identical gyroscopes with no coupling – effect on drive frequency on amplitude response (Baglio *et al.*, 2004b). Reproduced by kind permission of EUROSENSORS 2004

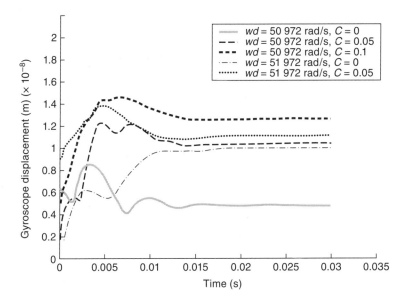

Figure 5.37 Effect of coupling on the variation of driving frequency (Baglio *et al.*, 2004b). Reproduced by kind permission of EUROSENSORS 2004

All of the gyroscopes synchronize again and oscillate at the same frequency and phase when a coupling signal is applied, even if they are forced by two frequencies that are different from the nominal one.

When the drive frequency is set to the mean resonance frequency for the non-identical array, the worst amplitude and synchronization performance is observed. This is caused by the phase-lag having the greatest variation at resonance.

5.6.1.1 Devices Realization

The layout of the described prototypes has been processed by the JAZZ semiconductor foundry, whereas etching of the devices has been performed at CMU (Carnegie Mellon University), by exploiting their facilities in performing isotropic *Silicon Deep Reactive Ion Etching* procedures.

The designed testing microprototypes have been mainly conceived to have a central plate suspended by compliant beams and electrically separated interdigitated comb fingers sets, to perform electrostatic actuation, electrostatic coupling and to detect plate displacements along the sense axis, as illustrated in Figure 5.38.

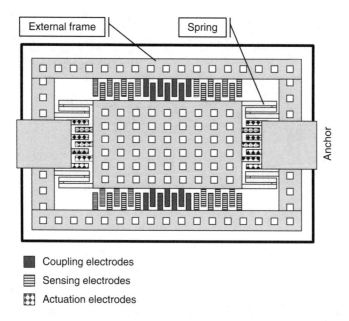

Figure 5.38 Schematic of the device prototype (Baglio *et al.*, 2004b). Reproduced by kind permission of EUROSENSORS 2004

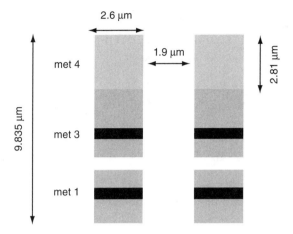

Figure 5.39 Comb fingers' cross-sections obtainable with four metal layers of the adopted technology (Baglio *et al.*, 2004b). Reproduced by kind permission of EUROSENSORS 2004

Structural design issues have included the realization of narrow gaps for actuation, wide structures for rigidity and wiring and a rigid frame anchored to the device symmetry axis ends. This arrangement has allowed the 'curl-matching' between fixed and moveable comb fingers to be improved.

To this purpose, a 'multi-level metal' has been used for routing signals, as well as for obtaining tall comb fingers, a 'stack' of about 10 μm, by using the fourth (top) metal layer as an etching mask as well as protection for eventual embedded electronics. In such a case, the minimum top metal width is 2.6 μm whereas the minimum gap (top metal) is 1.9 μm, as shown in Figure 5.39.

The device prototypes have been processed and released and some images obtained by optical microscopy and scanning electron microscopy (SEM) are reported in Figure 5.40 and Figures 5.41 and 5.42, respectively.

The characterization phase indicates that we should start forcing three oscillators along the device plane with three slightly different frequencies, through the actuation combs, sensing their displacements along the same axis through sense combs and then force coupling of the devices and phase synchronization through the coupling combs.

As concluding remarks on this latter case-study, some aspects are worth noting, as follows:

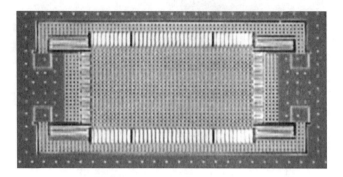

Figure 5.40 Micrograph of the testing device prototype, after releasing (Baglio *et al.*, 2004b). Reproduced by kind permission of EUROSENSORS 2004

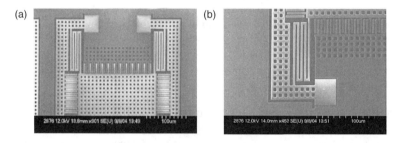

Figure 5.41 Scanning electron micrographs of a testing device prototype, showing magnifications of the anchoring area, springs and external frame (Baglio *et al.*, 2004b). Reproduced by kind permission of EUROSENSORS 2004

Figure 5.42 Scanning electron micrographs of a testing device prototype, showing magnification of the interdigitated fingers (Baglio *et al.*, 2004b). Reproduced by kind permission of EUROSENSORS 2004

- Coupling shows large benefits in compensating the effects of electromechanical parameters' spreading, usually occurring for micro-linear vibrating gyroscopes, and generally for micromachined MEMS, operated in the open-loop situation.

- Synchronization of the array, both in phase and frequency, allows us to improve system sensitivity, with respect to the operation of a single device, maintaining a simple electronic circuitry design which includes a sole demodulator for the array output.

ACKNOWLEDGEMENTS

1. Section 5.3.2.1 'The SOI micro-resonator has been... without affecting the exciter dynamic behaviour.', pp. 116–119. Portions of the text are reproduced from S. Baglio, V. Ferrari, A. Ghisla, V. Sacco, N. Savalli and A. Taroni (2005). SOI Mass-Sensitive Microresonators with Off-Chip Piezoelectric Excitation by PZT Thick Film, in *Proceedings of IMTC 2005, IEEE Instrumentation and Measurement Technical Conference*, Ottawa, ON, Canada, May 17–19, pp. 1736–1739, © 1996 IEEE and are reproduced by kind permission of IEEE.
2. Section 5.5 'Preliminary characterization of... realizing a kind of mass spectrometer.', pp. 127–130. Portions of the text are reproduced from S. Baglio, V. Ferrari, A. Ghisla, V. Sacco, N. Savalli and A. Taroni (2005). SOI Mass-Sensitive Microresonators with Off-Chip Piezoelectric Excitation by PZT Thick Film, in *Proceedings of IMTC 2005, IEEE Instrumentation and Measurement Technical Conference*, Ottawa, ON, Canada, May 17–19, pp. 1736–1739, © 1996 IEEE and are reproduced by kind permission of IEEE.
3. Section 5.6.1 'Apart from discussions regarding noise... demodulator for the array output.', pp. 134–145. Portions of text are reproduced from S. Baglio, A. Bulsara, J. Neff and N. Savalli (2004b). Array of coupled CMOS resonant vibrating gyroscopes, in *Proceedings of Eurosensors XVIII*, Rome, Italy, September 13–15, pp. 466–467 and are reproduced by kind permission of EUROSENSORS 2004.
4. Section 5.6 'Provided that there is a means... single or multi-degree of freedom system.', pp. 133–134. Portions of text are reproduced from W.A. Clark, R.T. Howe and R. Horowitz (1996). Surface micromachined z-axis vibratory rate gyroscope, in *Technical Digest, IEEE Solid-State Sensors and Actuators Workshop*, Hilton Head Island, SC, USA, June 1996, pp. 283–287, © 1996 IEEE and are reproduced by kind permission of IEEE.

REFERENCES

C. Acar and A. Shkel (2001). A Design Approach for Robustness Improvement of Rate Gyroscopes, in *MSM 2001, Technical Proceedings of the 2001 International Conference on Modeling and Simulations of Microsystems*, Hilton Head Island, SC, USA, March 19–21, pp. 80–83.

S.E. Alper and T. Akin (2002). A symmetric surface micromachined gyroscope with decoupled oscillation modes, *Sensors Actuators A: Phys.*, **97–98**, 347–358.

F. Ayazi and K. Najafi (1998). Design and fabrication of a high performance polysilicon vibrating ring gyroscope, in *Proceedings of the IEEE Micro Electro Mechanical Systems Workshop (MEMS'98)*, Heidelberg, Germany, February, pp. 621–626.

S. Baglio, M. Bloemer, N. Savalli and M. Scalora (2001). Development of Novel Opto-Electro-Mechanical Systems based on Transparent Metals PBG Structures, *IEEE Sensors J.*, 1, 288–295.

S. Baglio, S. Castorina, J. Esteve and N. Savalli (2004a). Highly sensitive silicon micro-*g* accelerometers with optical output, in *Proceeding of ISCAS 2004, IEEE International Symposium on Circuits and Systems*, Vancouver, BC, Canada, May 23–26, pp. 868–871.

S. Baglio, A. Bulsara, J. Neff and N. Savalli (2004b). Array of coupled CMOS resonant vibrating gyroscopes, in *Proceedings of Eurosensors XVIII*, Rome, Italy, September 13–15, pp. 466–467.

S. Baglio, V. Ferrari, A. Ghisla, V. Sacco, N. Savalli and A. Taroni (2005). SOI Mass-Sensitive Microresonators with Off-Chip Piezoelectric Excitation by PZT Thick Film, in *Proceedings of IMTC 2005, IEEE Instrumentation and Measurement Technical Conference*, Ottawa, ON, Canada, May 17–19, pp. 1736–1739.

W.A. Clark, R.T. Howe and R. Horowitz (1996). Surface micromachined z-axis vibratory rate gyroscope, in *Technical Digest, IEEE Solid-State Sensors and Actuators Workshop*, Hilton Head Island, SC, USA, June 1996, pp. 283–287.

C. Di Nucci, A. Fort, S. Rocchi, L. Tondi, V. Vignoli, F. Di Francesco and M.B. Serrano Santos (2003). A Measurement System for Odor Classification Based on the Dynamic Response of QCM Sensors, *IEEE Trans. Instrum. Meas.*, 52, 1079–1086.

G.K. Fedder (1997). Integrated MEMS in conventional CMOS, in *Proceedings of the NSF/ASME Workshop on Tribology Issues and Opportunities in MEMS*, Kluwer Academic Publishers, Dordecht, The Netherlands, pp. 2821–2824.

G.K. Fedder, S. Santhanam, M.L. Reed, S.C. Eagle, D.F. Guillou, M.S.C. Lu and L.R. Carley (2000). Laminated high-aspect ratio structures in a conventional CMOS process, *Sensors Actuators A: Phys.*, 57, 103–110.

V. Ferrari, D. Marioli and A. Taroni (1997). Thick-film resonant piezo-layers as new gravimetric sensors, *Meas. Sci. Technol.*, 8, 42–48.

V. Ferrari, D. Marioli and A. Taroni (2001). Theory, modelling and characterization of PZT-on-alumina resonant piezo-layers as acoustic-wave mass sensors, *Sensors Actuators A: Phys.*, 92 182–190.

D.W. Fogliatti (2002). Interconnected Resonant Vibratory Gyroscopes, in *Proceedings of ISCAS 2002, IV*, Scottsdale, AZ, USA, May 26–29, pp. 289–292.

J.M. Gere and S.P. Timoshenko (1997). *Mechanics of Materials*, 4th Edition, PWS, Boston, MS, USA, Ch. 9, pp. 599–680.

O. Gigan, H. Chen, O. Robert, S. Renard and F. Marty (2002). Fabrication and characterization of resonant SOI micromechanical silicon sensors based on DRIE micromachining, freestanding release process and silicon direct bonding in *Proceedings of SPIE*, vol. 4936, Nano and Microtechnology: Materials, Processes, Packaging, and Systems, Melbourne, Australia, pp. 194–204.

D.F. Guillou, S. Santhanam and L.R. Carley (2000). Laminated, sacrificial-poly MEMS technology in standard CMOS, *Sensors Actuators A: Phys.*, 85, 346–355.

M. Hashimoto, C. Cabuz, K. Minami and M. Esashi (1995). Silicon resonant angular rate sensor using electromagnetic excitation and capacitive detection, *J. Micromech. Microeng.*, 5, 219–225.

P. Hauptmann (1991). Resonant Sensors and Applications, *Sensors Actuators A: Phys.*, **25–27**, 371–377.

R.T. Howe (1987). Resonant Microsensor, in *Proceedings of the 4th International Conference on Solid-State Sensors anc Actuators (Tranducers'87)*, Tokyo, Japan, June 2–5, pp. 843–848.

J.W. Judy (2001). Microelectromechanical systems (MEMS): fabrication, design and applications, *Smart Mater. Struct.*, **10** 1115–1134.

H. Kawai, K. Atsuchi, M. Tamura and K. Ohwada (2001). High-resolution microgyroscope using vibratory motion, *Sensors Actuators A: Phys.*, **90** 153–159.

G.T.A. Kovacs (1998). *Micromachined Transducers Sourcebook*, McGraw-Hill,New York, NY, USA.

M. Kraft (2000). Micromachined inertial sensors: The state of the art and a look into the future, *IMC Meas. Cont.*, **33**, 164–168.

M.J. Madou (2002). *Fundamentals of Microfabrication: The Science of Miniaturization*, 2nd Edition, CRC Press, Boca Raton, FL, USA.

K. Maenaka and T. Shiozawa (1994). A study of silicon angular rate sensors using anisotropic etching technology, *Sensors Actuators A: Phys.*, **43**, 72–77.

K.W. Markus and D.A. Koester (1996). Multi-User MEMS Process (MUMPS) *Introduction and Design Rules*, MCNC Electronic Technology Division, Research Triangle Park, NC, USA.

V.M. Meccea (1994). Loaded vibrating quartz sensors, *Sensors Actuators A: Phys.*, **40**, 1–27.

Y. Mochida, M. Tamura and K. Ohwada (2000). A micromachined vibrating rate gyroscope with independent beams for the drive and detection modes, *Sensors Actuators A: Phys.*, **80**, 170–178.

Y. Oh, B. Lee, S. Baek, H. Kim, J. Kim, S. Kang and C. Song (1997). A surface-micromachined tunable vibratory gyroscope, in *Proceedings of the IEEE Micro Electro Mechanical Systems Workshop (MEMS'97)*, Japan, January, pp. 272–277.

J.A. Plaza, A. Liobera, J.Berganzo, J. Garcia, C. Dominguez and J. Esteve (2002). Stress Free Quad Beam Optical Silicon Accelerometers, in *Proceedings of IEEE Sensor 2002*, Orlando, FL, USA, June 11–14, pp. 1064–1068.

N. Savalli, S. Baglio, S. Castorina, V. Sacco and V. Ferrari (2004). Development of hybrid SOI-based microgravimetric sensors, in *Proceedings of SPIE International Symposium, Smart Structures and Materials*, Vol. 5389, San Diego, CA, USA, March 14–18, pp. 256–266.

A.A. Seshia, R.T. Howe and S. Montague (2002). An integrated microelectromechanical resonant output gyroscope, in *Proceedings of the 15th IEEE Micro Electro Mechanical Systems Conference*, Las Vegas, NV, USA, January 20–24, pp. 722–726.

A. Shkel, R. Horowitz, A.A. Seshia, P. Sungsu and R.T. Howe (1999). Dynamics and control of micromachined gyroscopes, in *Proceedings of the American Control Conference*, Vol. 3, June, pp. 2119–21124.

J. Soderkvist (1996). Micromachined vibrating gyroscopes, in *Proceedings of the SPIE 1996 Symposium on Micromachining and Microfabrication*, Austin, TX, USA, 2882, pp. 152–160.

J.B. Starr (1990). Squeeze Film Damping in Solid State, Accelerometer, in *Technical Digest, IEEE Solid-State Sensors and Actuators Workshop*, Hilton Head Island, SC, USA, June, pp. 44–47.

K. Tanaka, Y. Mochida, M. Sugimoto, K. Moriya, T. Hasegawa, K. Atsuchi and K. Ohwada (1995). A micromachined vibrating gyroscope, *Sensors Actuators A: Phys.*, **50**, 111–115.

N. Tas, T. Sonnenberg, H. Jansen, R. Legtenberg and M. Elwenspoek (1996). Stiction in Surface Micromachining, *J. Micromech. Microeng.*, **6**, 385–397.

H.A.C. Tilmans, M. Elwenspoek and J.H.J. Fluitman (1992). Micro resonant force gauges, *Sensors Actuators A: Phys.*, **30**, 35–53.

J.G. Webster (Ed.) (1999). *The Measurement, Instrumentation and Sensors Hanbook*, CRC Press, Boca Raton, FL, USA, pp. 33.1–3.11.

H. Xie and G.K. Fedder (2003). Fabrication, characterization and analysis of a DRIE CMOS-MEMS gyroscope, *IEEE Sensors J.*, **3**, 622–631.

H. Xie, L. Erdmann, X. Zhu, K. Gabriel and G. Fedder (2002). Post-CMOS Processing For High-aspect-ratio Integrated Silicon Microstructures, *J. Microelectromech. Syst.*, **11**, 93–101.

N. Yazdi, F. Ayazi and K. Najafi (1998). Micromachined Inertial Sensors, *Proc. IEEE*, **86**, 1640–1659.

G. Zhang, H. Xie, L.E. de Rosset and G. Fedder (1999). A lateral capacitive CMOS accelerometer with structural curl compensation, in *Proceedings of the 12th IEEE International Conference on Micro Electro Mechanical Systems (MEMS'99)*, Orlando, FL, USA, January, pp. 606–611.

6

Scaling of Energy Sources

6.1 INTRODUCTION

The analysis of scaling laws for microsensors, microactuators, and more generally microsystems, has regarded the way they operate and interact with the environment when their dimensions shrink. In the previous chapters, several scaling effects related to the actuation and sensing performance have been addressed in some concrete applications and examples. This chapter will be devoted to an analysis of the scaling effects on the energy source and the energy supply issues related to microsystems. In particular, the challenging application of an autonomous microsystem will be taken into account here for the purpose of a wider treatise on the scaling effects on energy sources.

The autonomous microsystem is intended here as a microsystem to which energy is supplied by means of an integrated energy source or by a suitable energy supply method or strategy which does not physically link the microsystem to a 'macroscopic' source (for example, wires connected to a power supply).

In the authors' vision, the autonomous microsystem, for excellence, is a *microrobot*. In fact a microrobot, in which actuators, sensors and electronics are integrated together into a unique, complex device, condenses a wide range of scaling issues and, among the most important ones, there is that of the energy supply strategy. From a general point of view, the problem is that of providing the required amount of energy

Scaling Issues and Design of MEMS   S. Baglio, S. Castorina and N. Savalli
© 2007 John Wiley & Sons, Ltd. Portions of text reproduced by kind permission of Elsevier

to a microrobot which moves, senses, interacts with its environment, processes and even transmits signals to a remote unit.

In this chapter, the problems related to the scaling of energy sources will be addressed by taking into account a hypothetical autonomous microrobot, and an energy supply strategy based on the exploitation of photothermal effects will be also presented.

A first consideration should be made on the energy needs of an autonomous microrobot. Among the components integrated in such an application, actuators are, in general, the ones characterized by the higher power consumption. In fact, as shown in Chapter 2, the energy needed by a given actuator is proportional to the amount of work it has to perform. The energy required for actuation may be more than one order of magnitude higher than that necessary for sensors and electronics. Therefore, the aspects relative to the supply of energy to microactuators are critical in the analysis and design of energy supply strategies for autonomous microsystems.

Among the various actuation methods, capacitive and thermal strategies are the most largely applied in standard IC technologies, mainly due to their intrinsic simplicity and low realization cost; in fact, both the actuation methods do not need any 'exotic' materials and have, generally, simple structures that can be realized with conventional technological steps and facilities.

With respect to capacitive electrostatic ones, thermal actuators generally allow larger displacements and can exert larger forces on a load, as shown in Chapter 2. On the other hand, thermal actuators are slower than electrostatic ones, because their dynamic is limited by the slow thermal exchange phenomena, but with miniaturization some improvements in terms of speed can be expected, However, the long response times are not necessarily a disadvantage in the sense that the application of thermal actuators is restricted to those fields where slow motions but large forces/displacements are required. This is the case for microrobotics and thus thermal microactuators are more suitable than capacitive ones.

The traditional 'macroscale' energy sources, solid-state power supplies and batteries, have some characteristics that make them unfeasible for such kinds of application. In fact, cables connecting a microrobot to a solid-state power supply may be too stiff, thus impeding the microrobot movements and/or limiting its operative range. Batteries, even the smallest types commercially available (as used in whistwatches) may be too big and heavy for a microrobot. Moreover, the operative life of a microrobot may be limited by the finite amount of energy stored in a battery, and replacing or recharging operations may not be feasible or convenient. Therefore, the only feasible and functional solution appears to be an external power supply source that allows the transferring of energy to the system without wires.

Several 'wireless' power supply strategies for microrobotics applications have been reported in the literature. Among the different methods, the transfer of energy through electromagnetic waves is one of the most effective. For example, Shibata *et al.*, (1998) presented a microwave-based energy supply systems for 'in-pipe' micromachines. One of the advantages of radiating energy over batteries, for example, is that the former scales as the surface area (the irradiated surface), while the energy stored in a battery scales as the volume.

Another approach exploits the mechanical vibrations produced by a vibrating floor to excite the legs of a small robot (Mayashi and Iwatsuki, 1998). However, this requires the microsystem to operate in a restricted, structured environment.

Integrated lithium microbatteries have been reported (West *et al.*, 2002), and also integrated photovoltaic cells (Lee *et al.*, 1995). However, both these approaches make use of non-standard IC materials and thus their integration could be difficult and expensive compared to other approaches.

The energy supply strategy that will be presented in this chapter exploits photothermomechanical and photothermoelectric energy conversions and relies on the use of focused light beams through integrated microlenses, to locally improve the energy density (Baglio *et al.*, 2002; Baglio *et al.*, 2003). The combined use of both systems allows supplying energy to the thermally actuated legs and the on-board electronics of a Si-MEMS microrobot.

6.2 ENERGY SUPPLY STRATEGIES FOR AUTONOMOUS MICROSYSTEMS

For the purpose of discussing the issues related to the scaling of energy sources, and to introduce a possible energy supply strategy for autonomous microsystems, the novel photothermomechanical actuation strategy will be addressed now. It will be supposed to apply such a strategy to a thermal actuator of the kind discussed in Chapter 2, a bilayer-like cantilever. A laser beam will be considered as the light source.

The light incident on the actuator surface heats up the cantilever that will bend due to the different coefficients of thermal expansion of the two layers. The degree of heating, and then the temperature reached by the structure and the amount of work performed by the actuator, depend on the power density of the light source and the exposure time. By controlling one or both of these source parameters, complete remote control of the actuators can be achieved.

It is worth noting that direct actuation with light can be achieved with photostrictive materials (Poosanaas *et al.*, 2000), but for the purpose of this work they are not a choice because such materials are generally non-compatible with silicon technologies; moreover, the bilayer thermal actuators are quite simple and diffused in integrated devices.

6.2.1 Use of microlenses in photothermomechanical actuation

In Chapter 2, the relation between the temperature variation in the cantilever and the supplied energy has been derived in equation (2.10), which is also reported here for convenience:

$$\frac{T - T_\infty}{T_0 - T_\infty} = \exp\left(-\frac{t}{\tau_T}\right) + \frac{q_h A_h}{hA_c (T_0 - T_\infty)} \left[1 - \exp\left(-\frac{t}{\tau_T}\right)\right] \qquad (6.1)$$

In equation (6.1), the product of the light source power density, q_h, and the irradiated device area A_h, is the energy transferred to the actuator. None of these parameters can be increased in order to improve the amount of energy intake without important drawbacks on the actuator. In fact, the irradiated area can be increased, at most, to the whole device area; otherwise, it would mean dealing with a larger device, thus loosing the advantages of scaling. On the other hand, the energy density of the source may be constrained by the physical integrity of the device or by the cost of the source itself.

If the need to achieve higher energy transfer for the actuation purpose, for example, contrasts with the system dimensions, physical integrity or cost, the only feasible way to improve energy transfer is to make use of a strategy which allows achieving locally higher energy densities, while keeping acceptable levels elsewhere on the device. If energy is transmitted through electromagnetic waves, such kinds of 'energy collectors' are represented by antennae. If energy is transferred through light, this collecting function can be performed by a lens.

The lens actually collects the 'low-intensity' light incident on its 'large' surface and selectively concentrates (focuses) it over a small area on the device, thus achieving a local improvement in the light intensity (energy density).

This gain in energy density is the basic idea and the main advantage of the proposed photothermomechanical power supply method; however, the strategy has many other important advantages, which are essential characteristics for a proper operation:

- *Energy density gain.* This allows using a lower-power light source, thus reducing the overall system cost and the risk of overheating critical system parts (the high temperature values and variations are required only on thermal actuators, for example).
- *Selective heating.* Selectivity and directionality achieved by using lenses can be exploited to maximize the actuator performance and efficiency by concentrating energy in critical areas of the device.
- *Optimization of energy transfer.* Lenses can be designed with a suitable gain in order to compensate for eventual energy losses due to undesired light reflection and absorption phenomena.
- *Simpler light control.* Thanks to the local focusing of light at the device level, there is no need for focusing and/or tracking mechanisms at the source level. Only a suitable switching or modulation strategy is required. In such a way, a unique source could be used to supply energy to multiple microsystems (microrobots), without the need to focus on, or track the position of a particular microsystem (microrobot).

The principle of photothermomechanical actuation is represented in Figure 6.1, which shows a bilayer cantilever operating as a thermal actuator, with a micromachined lens placed over it that collects and focuses light over its surface.

The feasibility of microlenses through processes and materials compatible with silicon technology has been demonstrated (Sankur *et al.*, 1995; Erdmann and Efferenn, 1997; Fujita and Toshiyoshi, 1997). The realization of a microlens is schematically reported in Figure 6.2; the lens pattern is first reproduced in a circular photoresist 'island' deposited over a given substrate, through a low-temperature (180–200 °C) thermal process that causes the 'reflow' of the photoresist; then, a selective *reactive ion etching* process transfers the pattern to the substrate. A scanning electron micrograph of a microlens fabricated in such a way is reported in Figure 6.3; here, the substrate

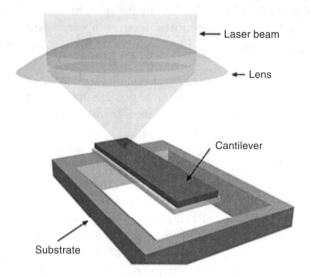

Figure 6.1 Structure of the proposed thermomechanical microactuator. The lens focuses a laser beam onto the cantilever, so causing it to heat. Due to the different coefficients of thermal expansion of the layered materials, the cantilever will bend

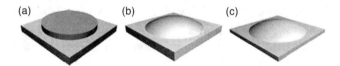

Figure 6.2 Schematic fabrication process of the microlens: (a) a circular photoresist 'island' is realized on a substrate; (b) photoresist thermal 'reflow'; (c) the lens pattern is transferred onto the substrate after a selective reactive ion etching process

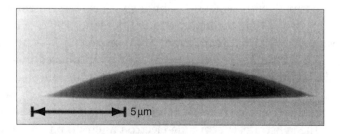

Figure 6.3 Scanning electron micrograph of a silicon microlens sample

is silicon – however, the same process can be used for other substrates, like silicon dioxide or silicon nitride, to realize lenses for the visiblelight range.

In Chapter 2, the thermomechanical behaviour of the thermal microactuator discussed here has been examined. Now such an analysis will be expanded in order to take into account the effects of the lens and to derive the design equations and some scaling considerations.

The device described in Chapter 2 will be supposed to receive actuation energy from a light source, for example, a laser beam and then the effects of the lens will be discussed.

In the following analysis, it will be supposed that a light source irradiates energy uniformly, at least in a given region of space, with q_s the energy density and A_L the area of the lens convex surface. Moreover, the lens will be supposed to be completely transparent to the wavelengths of interest. This latter assumption does not affect the generality of the discussion because the lens gain can be adjusted in order to compensate for eventual energy losses. Under these hypotheses, conservation of energy through the lens gives:

$$q_s A_L = q_h A_h \text{ or } q_h = \frac{A_L}{A_h} q_s = G_L q_s \qquad (6.2)$$

where q_h and A_h are the energy density and the area of the focused light spot, respectively.

From equation (6.2), the lens 'amplifies' the energy density of the source by a factor $G_L = A_L/A_h$, which is the gain of the lens. Equation (6.2) substituted in equation (6.1), gives:

$$\frac{T - T_\infty}{T_0 - T_\infty} = \exp\left(-\frac{t}{\tau_T}\right) + \frac{q_s A_L}{h A_c (T_0 - T_\infty)} \left[1 - \exp\left(-\frac{t}{\tau_T}\right)\right] \qquad (6.3)$$

which shows that the use of lenses allow locally achieving the same temperature variation with a G_L times lower energy density from the source.

The lens has the shape of a spherical segment, as shown in Figure 6.4 with its characteristic dimensions: the radius of the sphere, r, the angle at the center, θ, the lens diameter, d and the 'sag height', H.

The convex surface area, A_L of the lens shown in Figure 6.4 is expressed as:

$$A_L = 2\pi r^2 (1 - \cos\theta) \qquad (6.4)$$

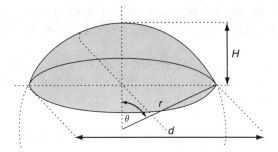

Figure 6.4 The geometry and dimensions of a spherical lens

However, equation (6.4) is almost useless in its actual form because the technology for the fabrication of microlenses does not allow easy control of the the parameters r and θ and thus it is more convenient to express the surface area of the lens, A_L, in terms of its diameter, d and its sag height, H. By simple geometrical consideration it follows that:

$$A_L = \pi \left(H^2 + \frac{d^2}{4} \right) \tag{6.5}$$

The dimensions and shape of a lens also determine a fundamental parameter, i.e. the focal length, f, which in turn affects the distance from the device at which the lens should be placed. In the socalled 'paraxial approximation' (Bahaa *et al.*, 1991), the focal length, f, is expressed as:

$$f = \frac{r}{n-1} \tag{6.6}$$

where n is the refraction index of the lens. From equations (6.4) and (6.5), this results in the following:

$$f = \left(\frac{1}{n-1} \right) \left(\frac{A_L}{2\pi H} \right) \tag{6.7}$$

Equations (6.3), (6.5) and (6.7) are *design equations* for the system considered here. The design flow starts from the specification on the desired actuator displacement, δ, which, as shown in Chapter 2, allows determining the required temperature variation, ΔT, subject to the constraint of the maximum allowable temperature. Equation (6.3) allows calculating the required heat flux on the cantilever once ΔT is known. Then, by using equations (6.5) and (6.7) the parameters of the lens can be calculated and the energy density of the light source can be determined.

6.2.2 Technologies, materials and design of photothermomechanical actuators

The realization of a thermal actuator prototype of the type discussed here has already been addressed in Chapter 2. A standard CMOS technology, together with a bulk anisotropic etching process (Baglio *et al.*, 1999) have been taken into account there to realize a bilayer-type cantilever beam as the thermal actuator. The realization of the microlens over the thermal actuator requires additional processes and materials. In fact, lenses may be fabricated on additional layers grown or deposited over the standard CMOS ones, or alternatively they can be fabricated on a separate substrate and then assembled over the actuators by means of suitable supports/spacers. A schematic of such a kind of structure is shown in Figure 6.5.

The design of the thermal actuator prototype addressed in Chapter 2 will be revisited here by introducing photothermo mechanical actuation and energy supply strategy. The technology of reference is a standard CMOS one whose materials are reported in Table 2.1 (see Chapter 2), together with some significant physical properties. With reference to the structure reported in Figure 6.5, the cantilever is supposed to be built by the superposition of the DIFF and CONT oxides and the two metal layers and has dimensions of $300\,\mu m \times 44\,\mu m$. Such a structure has mass $M = 9.075 \times 10^{-11}\,kg$ and stiffness $K = 0.596\,N/m$.

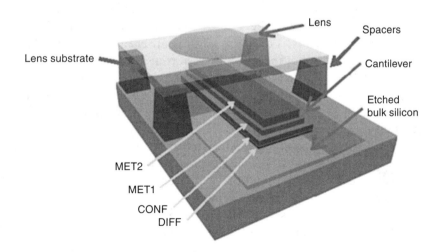

Figure 6.5 The structure of the photothermomechanical microactuator

The maximum operating temperature is limited by the melting point of the metal layers (Al alloys), which is about 730 K; therefore, the maximum, theoretical temperature variation is about 350 K. In these conditions, the surface and fluid temperatures are, respectively, $T_s = 650$ K and $T_f = 475$ K. From the thermophysical properties tabulated in Incropera and De Witt (2001), the Rayleigh number results in the following:

$$Ra = \frac{g\beta\,(T_s - T_\infty)\,L_c^3}{\nu\alpha} = 1.97 \times 10^{-4} \qquad (6.8)$$

where g is the gravitational field, β the volumetric thermal expansion coefficient, ν the viscosity, α the thermal diffusivity and L_c the characteristic length of the cantilever (defined as the ratio of the volume to the surface area). Due to the small value of the Rayleigh number, the following correlation will be applied for the Nusselt number:

$$Nu = 0.54 Ra^{0.25} \qquad (6.9)$$

where the Nusselt number is defined as:

$$Nu = \frac{h L_c}{k_f} \qquad (6.10)$$

with k_f being the thermal conductivity of the fluid. The resulting value for h is 67.6 W/m²/K.

In Chapter 2, it has been shown that the Biot number, Bi, is much smaller than unity and then the 'lumped' thermal capacitance model can be applied for the analysis of thermal transient. From that analysis, the thermal time-constant of the structure results in $\tau_T = 42$ ms. It has been calculated that the actuation energy density required by the structure is $q_h = 5.5 \times 10^6$ W/m² over a spot area A_h equal to the cantilever width. With these values, the heating time is about 50 ms, and then the actuation power results in $P = q_h A_h = 8.4$ mW.

The actuation power has been calculated under the assumption that all of the energy carried by the light and focused on the actuator is converted into heat. Clearly, this assumption does not hold if the materials used to realize the actuator have non-zero reflectivity. This is, for example, the case of the structure shown in Figure 6.5, where light is focused on a metal layer. The non-zero reflectivity of the material means losses in terms of energy efficiency; however, with a suitable design of the lens, its gain can partially or fully compensate these losses.

The design of lenses will be addressed in the following. In Figure 6.5 it is shown that the lens is placed at a given distance from the actuator through some spacers. Depending on available materials and technology, and on device dimensions, such kinds of spacers could be realized in different ways, for example, by means of some thick-film and micromachining processes. However, regardless of the assembly technique, some design considerations on the lens parameters will be discussed here.

The lens focuses light into a spot over the device surface. Ideally, this spot should be 'punctual', or with a very small diameter. In practice, the diameter of the focused spot should be slightly bigger than the actuator width in order to accommodate for alignment tolerances. Moreover, the lens area must be as large as higher is the gain of energy density that is required. These considerations constrain the design of the lens parameters, which are shown in Figure 6.6.

From Figure 6.6, it follows that:

$$f = \frac{db}{d - 2a} \tag{6.11}$$

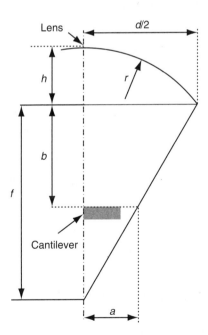

Figure 6.6 Schematic of the lens parameters of interest for the design

Since r must be greater than $d/2$, and f is given by equation (6.6), it follows that:

$$b > \frac{d - 2a}{2(n-1)} \qquad (6.12)$$

If the lens is made of SiO_2, which has $n = 1.5$, equation (6.12) is reduced to:

$$d < b + 2a \qquad (6.13)$$

In order to achieve the highest gain, and then the largest possible lens area, the lens diameter, d, should be as high as possible; therefore, the following design criterion will be adopted:

$$d = 0.9(b + 2a) \qquad (6.14)$$

Once b and d are known, the focal length, f, can be calculated from equation (6.11); therefore, the lens radius of curvature, r, can be derived from equation (6.6). The 'lens dome height', H, follows from the geometrical considerations given in Figure 6.6:

$$H = r - \sqrt{r^2 - \left(\frac{d}{2}\right)^2} \qquad (6.15)$$

Finally, the lens area is given by equation (6.5), and the 'gain' factor can be defined as the ratio between the lens area, A_L and the spot area, πa^2.

The calculated lens parameters, i.e. the focal length, f, the diameter, d and the dome height, H, are used at the lens fabrication stage to tune the process parameters, as reported in Sankur *et al.* (1995), Erdmann and Efferenn (1997) and Fujita and Toshiyoshi (1997).

6.3 PHOTOTHERMOMECHANICAL AND PHOTOTHERMOELECTRIC STRATEGIES FOR HIGHLY EFFICIENT POWER SUPPLY OF AUTONOMOUS MICROSYSTEMS

The photothermomechanical actuation strategy provides the high energy amount needed by actuators, as described in the preceding sections of this chapter. A complete autonomous microsystem (like a microrobot, for example) also needs 'on-board' electronics for control and signal-processing purposes. Power requirements for such electronics are usually smaller than those of actuators, especially if power electronics stages are not required, with the source of actuation being 'non-electric' as in the case of photothermomechanical actuators. In this section, it will be shown how the proposed energy supply system can be extended at very low cost to generate an electric power to supply some low power 'on-board' electronic systems.

The focused light beams used for photothermomechanical actuation produces local heating in the device structure and therefore non-zero temperature gradients may exist in selected areas of the device; these temperature gradients can be exploited to generate electric power if integrated thermoelectric elements are realized between the hottest and the coldest points of the device.

The use of microthermoelectric elements as thermal sensors has already been discussed in Chapter 3; here, they will be treated under the power-generation point of view. This is the basic idea of a complete photothermal energy supply system in which the heat generated by focused light beams is directly exploited to produce mechanical work in actuators and electric power in thermoelectric elements, leading to a high overall energy efficiency.

6.3.1 Photothermoelectric power generation

The most straightforward method to convert light into electric energy is to exploit the photovoltaic effect, and then by making use of photo-voltaic cells in microsystems. Microphotovoltaic devices have been reported in the literature (see Lee *et al.*, 1995, for example); however, the realization of such kinds of devices requires non-standard materials and processes, which is not in agreement with the design constraint of making use of standard CMOS technologies, adopted throughout this work – thus, they will not be taken into account here. Beyond the technology and cost concerns, the purpose of a photothermoelectric strategy is to extend and complete the capabilities of the photothermomechanical actuation method in a convenient and efficient way. In fact, as for application as thermal sensors, thermoelectric elements can be integrated at a very low cost and with a small design effort in almost any IC technology with no use of 'exotic' materials and processes.

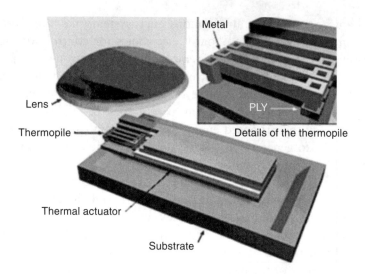

Figure 6.7 Illustration of a combined photothermomechanical and photothermo-electric device

The basic idea of coupling the photothermal conversion system with integrated thermoelectric generators (Baltes *et al.*, 1998; Strasser *et al.*, 2002) for the 'on-chip' generation of electric power, which supplies energy to the electronics, is schematically shown in Figure 6.7. As for the actuators, the efficiency of a thermoelectric generator is directly related to the thermal energy density; therefore, the use of the microlenses will improve the efficiency of the whole energy supply system and will allow the generation of higher electric power.

In a standard CMOS technology, as well as in any IC technologies, several conducting materials are available (metals and poly and crystalline silicon with various types and levels of doping). Each 'coupling' of such different materials is potentially a thermocouple; therefore, an integrated thermoelectric generator or a thermopile can be realized by means of a series connection of N thermocouples.

The open-circuit voltage, V_O, across a thermopile depends on the temperature difference, ΔT, between the hot and the cold junctions, through the relative Seebeck coefficient α (Baltes *et al.*, 1998; Strasser *et al.*, 2002), as expressed in equation (3.2) (see Chapter 3) which is reported here for convenience:

$$V_O = N\alpha\Delta T \qquad (6.16)$$

Given R_O, the series electrical resistance of the thermopile, the maximum electrical power is delivered by the thermoelectric generator when its load equals R_O, and is:

$$P_{MAX} = \frac{(N\alpha\Delta T)^2}{4R_O} \qquad (6.17)$$

Equation (6.17) can be rewritten in terms of the electrical resistivities, ρ_1 and ρ_2, of the materials used in the thermo-elements; moreover, with no loss of generality, the thermo-elements' dimensions can be assumed to be the same and thus the length l and cross-sectional area A:

$$P_{MAX} = \frac{NA\alpha^2\Delta T^2}{4l\rho} \qquad (6.18)$$

where ρ is the equivalent electrical resistivity of a thermocouple: $\rho = \rho_1 + \rho_2$.

In equation (6.18), N, A and l are design parameters, while a and ρ are process parameters. The temperature difference is a function of the incident power; therefore, a thermal analysis should be performed to determine this relation. The thermal analysis can be restricted to a single thermocouple. This is a parallel connection of two thermal conductors across which a temperature gradient ΔT exists, which produces a heat flow rate Q through them. To a first-order approximation, heat losses along the thermoelectric elements will be neglected. This assumption is reasonable because the elements should be realized on a insulating layer to avoid electrical short circuits; this insulating layer may also help in reducing thermal losses. The thermal problem to be analysed here is illustrated in Figure 6.8.

In the following analysis, the 'cold-junction' of the thermocouple will be supposed to be thermally connected to the substrate (at least, through a low-thermal-resistance path). The thermal mass of the substrate is much greater than that of the thermocouple; therefore, it can be viewed as a 'heat-sink' whose temperature (the temperature of the cold junction) is maintained constant at T_0. Actually, T_0 is slightly bigger than the ambient temperature and, in fact, the substrate may experience a certain heating – however, this effect could be neglected.

The heat rate, Q, propagates through the thermo-elements due to thermal conduction; therefore, in terms of the heat flux density, $q = Q/A$, is:

$$q = -k\frac{dT}{dx} \qquad (6.19)$$

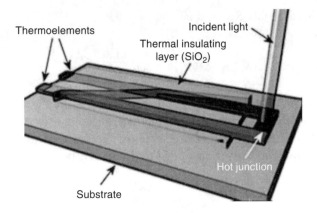

Figure 6.8   Three-dimensional illustration of a single thermocouple

where k is the equivalent thermal conductivity. The heat flux, q, is directly related to the energy density of the light source and, in the case of a pulsed light source (that is, in the most practical case), it can be viewed as the root-mean-square value of the pulsed heat flux. Equation (6.19), solved for the temperature difference ΔT (in terms of finite variations), gives:

$$\Delta T = \frac{ql}{k} \tag{6.20}$$

From the heat-conduction point of view, the two thermo-elements of a thermocouple are a parallel connection; therefore, the equivalent thermal conductivity can be defined as $k_1 + k_2$.

Equation (6.20), substituted into equation (6.18), gives the relation between the maximum power deliverable by the thermoelectric generator, P_{MAX}, and the energy density of the light source, q, as:

$$P_{MAX} = \frac{NAl\alpha^2 q^2}{4\left(\rho_1 + \rho_2\right)\left(k_1 + k_2\right)^2} \tag{6.21}$$

Equation (6.21) shows that the energy-conversion method has an intrinsic high efficiency because the electrical power generated depends on the square power of the incoming energy density.

Further improvements in terms of energy efficiency are achieved by introducing microlenses. As previously shown, the lens collects the energy coming from the light source over a large surface area and concentrates it over a small spot, so introducing a gain in terms of energy density, which is given by the ratio between the convex surface area of the lens and the area of the spot where the light is focused, that is:

$$q = q_s \frac{A_L}{A_H} \tag{6.22}$$

where q_s is the energy density of the light, A_L is the area of the convex surface of the lens and A_H is the area of the spot where light is focused. In the previous analysis, the lens is considered as an ideal optical element, without losses in the transfer of energy. The introduction of the gain of the lens, $G = A_L/A_H$, in equation (6.21) gives:

$$P_{MAX} = \frac{N A l \alpha^2 G^2 q_s^2}{4 (\rho_1 + \rho_2) (k_1 + k_2)^2} \tag{6.23}$$

In equation (6.23), it is shown that the gain, G, of the lens enters to the second power in the calculation of the maximum power deliverable by the photothermoelectric generator. It is also worth noting that this approach produces thermoelectric power without an appreciable overall heating of the system; in fact, the use of microlenses allows achieving the required energy density only in selected areas of the system.

Equation (6.23) can be rewritten in terms of power per unit of volume, $p = P_{MAX}/V$, resulting in a useful equation to compare the power supply system described here to other systems presented in the literature. Since the volume of the thermoelectric generator can be expressed as $V = NAl$, this results in:

$$p = \frac{\alpha^2 G^2 q_s^2}{4 (\rho_1 + \rho_2) (k_1 + k_2)^2} \tag{6.24}$$

The design of thermopiles in standard CMOS technology has already been addressed in Chapter 3, where the application of these devices as thermal sensors has been discussed. There, the technology, the design issues and the experimental verification of some prototypes have been reported. In application as sensors, the parameters of interest are, for example, the sensitivity, the resolution and the noise level, whereas in their use as thermoelectric generators, the power density and losses are of major concerns. Despite these differences, most of the design considerations made in Chapter 3 still hold; thus, the reader may refer to this chapter.

6.4   EFFICIENCY OF THE COMBINED ENERGY SUPPLY STRATEGY

For the sake of further proof of the efficiency of a combined photothermomechanical and photothermoelectric energy supply system this will be compared with a more conventional approach. In particular, in Sakakibara *et al.* (2002) a power-supply system based on an array of 100 integrated photovoltaic cells is presented. It is reported that such a system can deliver up to 8.9 mW with an incident light power of 1 W. The power per unit of volume is about 1.8×10^{10} W/m^3.

In the same conditions of energy density of the light source, the photothermoelectric generator proposed here delivers an electrical power density of $p = 25.8\,G^2$, derived from equations 6.23 and 6.24. If the lens has the following features: $a = w/2 = 22\,\mu$m and $b = 170\,\mu$m, according to symbols represented in Figure 6.6, then $d = 192.6\,\mu$m from equation 6.14, and it will introduce a gain factor G of about 25.8.

Therefore the p results: $p = 17.2 \times 10^3$ W/m^3. This value is smaller than that obtained by the photovoltaic device; however, for the sake of completeness, this comparison should also include the power delivered to the actuators by the photothermomechanical system, which amounts to 8.4 mW per actuation cycle. Thus, the advantage of the proposed power supply strategy over other systems will arise when combination of the photothermomechanical actuation and photothermoelectric power generation is considered. Moreover, the strategy presented here can be implemented in fully compatible CMOS technology, with a significant advantage in terms of costs over different power supply methods.

To further highlight the importance of a remote power supply strategy for autonomous microsystems, like the photothermomechanical system presented here, a comparison of the performance of such a system with those achievable by using hypothetical scaled batteries will be made here.

One of the smallest batteries commercially available is the CR1025 type (Energizer). The characteristics of such a battery are as follows:

- diameter, $d = 10.0$ mm
- height, $h = 2.5$ mm
- volume, $V_b = 196.25 \times 10^{-9}$ m^3
- weight, $w_b = 0.7$ g
- average mass density, $\rho_b = 3567$ kg/m^3
- Capacity, $C = 30$ mAh at 2 V
- stored energy, $E_b = 216$ J
- energy density, $\varepsilon_b = 1.1 \times 10^9$ J/m^3.

From Chapter 2 (Section 2.1.7), the characteristics of the thermal actuator are as follows:

- actuation power, $P_a = 8.4\,\text{mW}$
- heating time, $t_{on} = 50\,\text{ms}$
- actuation energy (per cycle), $E_a = 420\,\mu\text{J}$
- volume, $V_a = 1.5 \times 10^{-13}\,\text{m}^3$
- weight, $w_a = 35 \times 10^{-15}\,\text{kg}$.

With its original dimensions, the battery considered here would have enough stored energy to perform $E_b/E_a = 514 \times 10^3$ actuation cycles on a single actuator. The same number must be divided by the number of actuators if the same battery supplies multiple devices. However, such a number of actuation cycles could be achieved by using a battery which is $w_b/w_a = 2 \times 10^{10}$ times heavier than the actuator (thus making unfeasible the use of such a battery in a microrobot, for example).

It will be supposed now that the battery is hypothetically scaled to dimensions comparable with those of the actuator, by maintaining unaltered its energy density and electromotive force. For example, if the diameter of the battery is scaled down to $500\,\mu\text{m}$, thus 20 times smaller than the original value (the actuator length is $300\,\mu\text{m}$), the battery volume scales by a factor 20^3, then, the volume of the scaled battery will be $V_b^* = 24.5 \times 10^{-12}\,\text{m}^3$. Therefore, the energy stored in the scaled battery is $\varepsilon_b V_b^* = 27\,\text{mJ}$, which will be enough for only 64 actuation cycles on a single actuator; furthermore, the weight of the scaled battery would still result as 2.5×10^6 times higher than the weight of the actuator.

The example presented here clearly shows how batteries are ineffective if autonomous microsystems with 'on-board' actuators are taken into account. The actual small batteries are still too heavy for such microsystems, while hypothetical-scaled batteries can store too low a energy for the system to be really autonomous. Thus, remote, wireless power supply sources must be used in such applications.

REFERENCES

S. Baglio, L. Latorre and P. Nouet (1999). Resonant magnetic field microsensors in standard CMOS technology, IMTC/99, *Proc. of the 16th IEEE Instr. and Meas. Tech. Conf.*, Venice, Italy, May 24–26, pp. 452–457.

S. Baglio, S. Castorina, L. Fortuna and N. Savalli (2002). Modeling and design of novel photo-thermo-mechanical microactuators, *Sens. and Act. A: Phys.*, **101**, 185–193.

S. Baglio, S. Castorina, L. Fortuna and N. Savalli (2003). Highly efficient power supply strategy for autonomous microsystems, in *Proceedings of the International Advanced Robotics Program (IARP), Workshop on Micro Robots, Micro Machines and Micro Systems*, Moscow, Russia, April 24–25.

H. Baltes, O. Paul and O. Brand (1998). Micromachined thermally based CMOS microsensors, *Proceedings of the IEEE*, 86, 1660–1668.

Energizer. Website [www.energizer.com].

L. Erdmann and D. Efferenn (1997). Technique for monolithic fabrication of silicon microlenses with selectable rim angles, *Opt. Eng.*, 36, 1094–1098.

H. Fujita and H. Toshiyoshi (1997). Micro-optical devices, in *Handbook of Microlithography, Micromachining and Microfabrication*, Vol. 2, *Micromachining and Microfabrication*, P.R. Choudhury (Ed.), SPIE, The International Society of Optical Engineering, Washington, DC, USA, pp. 435–516.

J.B. Lee, Z. Chen, M.G. Allen, A. Rohatgi and R. Arya (1995). A miniaturized high voltage solar cell array as an electrostatic MEMS power supply, *IEEE J. Microelectromech. Syst.*, 4, 102–108.

M. Mayashi and N. Iwatsuki (1998). Micro moving robotics, in *Proceedings of the International Symposium on Micromechatronics and Human Science (MHS)*, Nagoya Congress Centre, Nagoya, Japan, November 25–28, pp. 41–50.

P. Poosanaas, K. Tonooka and K. Uchino (2000). Photostrictive actuators, *Mechatronics*, 10, 467–487.

T. Sakakibara, H. Izua, T. Shibata, H. Tarui, K. Shibata, K. Kiyama and N. Kawahara (2002). Multi-source power supply system using micro-photovoltaic devices combined with microwave antenna, *Sens. and Act. A: Phys.*, 95, 208–211.

B.E.A. Saleh and M.C. Teich (1991). *Fundamentals of Photonics*, John Wiley & Sons, Inc., New York, USA.

H. Sankur, E. Motamedi, R. Hall, N.J. Gunning and M. Khoshnevisan (1995). Fabrication of refractive microlens array, in *Micro-Optics/Micromechanics and Laser Scanning and Shaping*, Proceedings of the SPIE, Vol. 2383, M.E. Motamedi and L. Beiser (Eds), SPIE, The International Society of Optical Engineering, Washington, DC, USA, pp. 179–183.

T. Shibata, T. Sasaya and N. Kawara (1998). Microwave energy supply system for in-pipe micromachine, in *Proceedings of the International Symposium on Micromechatronics and Human Science (MHS)*, Nagoya Congress Centre, Nagoya, Japan, November 25–28, pp. 237–242.

M. Strasser, R. Aigner, M. Franosch and G. Wachutka (2002). Miniaturized thermoelectric generators based on polySi and polySi–Ge surface micromachining, *Sensors Actuators A: Phys.*, 97–98, 535–542.

W.C. West, J.F. Whitacre, V. White and B.V Ratnakumar (2002). Fabrication and testing of all solid-state microscale lithium batteries for microspacecraft applications, *J. Micromech. Microeng.*, 12, 58–62.

7

Technologies and Architectures for Autonomous MEMS Microrobots

7.1 DESIGN ISSUES IN MICROROBOTS

In the previous chapters, several types of microsystems have been analysed and discussed by taking into account the effects of scaling on their operation, design and optimization. These devices include microactuators, microsensors and energy sources; electronics for the control and/or signal processing is also included, but it is not the aim of this work to address scaling issues for electronic devices.

Each of the devices analysed previously, or presented in the literature, may potentially operate as standalone systems, in array configurations or in combination with different types of devices in order to implement more complex functions (for example, different types of sensors and/or sensors and actuators, power supply systems). In other words, such devices are potential building blocks for higher-complexity systems.

By extrapolating the complexity of the systems which can potentially be realized by suitable choice of building blocks, the most complex one is a microrobot. In fact, a microrobot, which is autonomous both from the power supply and the control/processing points of view, which can move itself in a given environment and interact (sense and actuate) with it in performing well-defined tasks, groups in itself the wider spectrum

Scaling Issues and Design of MEMS   S. Baglio, S. Castorina and N. Savalli
© 2007 John Wiley & Sons, Ltd

of design, fabrication, validation and, most importantly, scaling issues related to microsystems.

Beyond the analysis, design, fabrication and testing efforts required by a single device, or class of devices, the closed integration of electronics, actuators, sensors and power supply in a microrobot is challenging from many points of view. To cite a few:

- Technology
- Structure
- Functionality
- Cost.

In detail, a hypothetical microrobot is designed to perform given tasks; thus, it must satisfy a set of specifications, which can be relative to several aspects of the microrobot's functions: like type and number of sensing capabilities and characteristics; type and number of actuation functions, like locomotion and interaction with the surroundings, speed and operating range; processing, control and communications capabilities; autonomy, energy budget; characteristics of the operative environment (type of surfaces, corrosive substances, flammable or explosive compounds, materials, accessibility, etc.); cost of the overall system.

For a given function, that is, sensing, actuation, electronics, in most cases the relative specifications can be satisfied with stand-alone solutions. However, the assembly or integration of single, operating and optimized structures in a unique system is, in general, not straightforward and does not guarantee the feasibility and functionality of the final structure. Novel issues arise when a complex structure like a microrobot is concerned and a careful analysis of design choices must be performed.

The specifications given for the design of a microrobot, for the single functions or class of functions, may lead to the identification of some design solutions having a given design space; thus, it may make sense to search the optimal 'overall' design solution in the 'intersection' of these design spaces. An important characteristic that design solutions should present as a common element is the technology; individual parts and functions of the microrobot should be preferably realized with the same technology or compatible ones, in order to minimize the effects of interactions, contaminations and incompatibilities due to the use of different materials and processes. Moreover, the use of compatible technologies minimizes fabrication cost.

Another characteristic that should be common in the design space is the structural feasibility and compatibility of the single functions. In particular, the choice of a given structure for a device, say an actuator, must not interfere with the operation of the other structures in the microrobot. The nature of these interferences could be mechanical, thermal, electrical, electromagnetic and optical. Thus, even if the single design solutions are technologically compatible, the designer should also check the structural, logical and functional compatibility.

The final microrobot assembly must guarantee the full functionality of each element and the overall functionality of the system, in order to satisfy the required specifications in performing the required tasks.

A fundamental driver in the whole design process is the cost, which, depending on the specific application and specifications, influences the design choices at every stage.

The ultimate issue in realizing a microrobot is the choice of a given set of design solutions (i.e. sensors, actuators, electronics and power supply) which can satisfy the specifications, which can be realized and integrated with a unique technology or with fully compatible processes, which can be realized and assembled in a final structure preserving feasibility and functionality of both the single elements and the overall system and, last but not least, which can be implemented at a reasonable cost.

7.2 A MICROROBOT ARCHITECTURE BASED ON PHOTOTHERMAL STRATEGY

An example of microrobot architecture in which the previous paragraph's considerations are applied will be given here. The specifications are relative to the actuation and energy-supply strategy of the microrobot and the proposed architecture makes use of the photothermomechanical and photothermoelectric actuation and energy-supply strategy discussed in the previous chapters. In particular, the leading idea of the microrobot architecture is to make use of the photothermomechanical actuation strategy to implement thermal actuators for the microrobot locomotion; in fact, thermal actuators can produce the large forces/strokes required for locomotion on the microscale and the photothermal excitation can provide the required high actuation energy while preserving the autonomy of the system, where batteries and cables fail. It has been shown that photothermomechanical actuators can be fabricated in full compatible CMOS technologies, which is a fundamental characteristic with respect to the discussed design considerations. The use of full CMOS compatible processes allows for the 'safe' integration of mechanics and electronics.

In the proposed architecture, the thermal actuators are part, or the whole legs of the microrobot. The thermal actuator examined in Chapter 2 has a planar structure, whereas the legs of a microrobot should have a more complex, three-dimensional structure, allowing sustain ability and locomotion. Examples of three-dimensional structures achieved from planar (two-dimensional) ones by making use

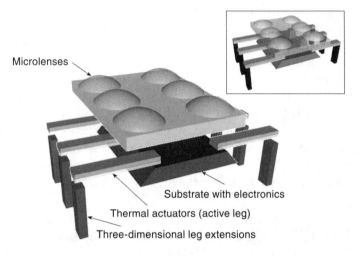

Microlenses

Substrate with electronics
Thermal actuators (active leg)
Three-dimensional leg extensions

Figure 7.1 Exploded view of the proposed microrobot structure. The inset shows a complete view of the device

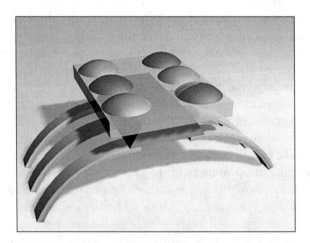

Figure 7.2 Three-dimensional illustration of the complete microrobot

of micromachined hinges (Yeh and Pister, 2000) or folding techniques (Ebefors *et al.*, 1999) have been reported in the literature.

Microrobot electronics for signal processing, control, etc. can be integrated in the same substrate of the microactuators and/or in another substrate, assembled with the other in a late phase of the robot realization. Microlenses must be realized in a transparent substrate or layer, and placed on top of the final device. In fact, the realization of microlenses does not rely on standard CMOS steps; however, they can be fabricated in a thick transparent layer or in a separate substrate and successively assembled into the rest of the structure. Both of the solutions show potential to be implemented in a compatible way.

The hypothesized microrobot architecture is illustrated in Figures 7.1 and 7.2.

7.3 A MICROROBOT AS A PARADIGM FOR THE ANALYSIS OF SCALING IN MICROSYSTEMS

The evolution of microelectronic devices has been characterized by the scaling of their characteristic feature size towards smaller dimensions. The reason for such a scaling trend is the continuous research into better processing capabilities, which means a higher number of smaller transistors on the chip. Integrated circuits with typical feature sizes in the submicrometre range are currently fabricated and commercialized, and research is pushing such feature sizes into the regime of few tens of nanometres and even smaller.

The huge potential for low-cost and large-scale fabrication of the semiconductor-microelectronics technologies has represented a very powerful and flexible platform for the conception and realization of miniaturized micromechanical structures, opening the way to the realization of miniaturized sensors and actuators.

The basic idea of MEMS is to exploit the same technology processes of the microelectronics industry to realize, on the same chip, both the electronics and the functionalized micromechanical structures which represent the interface with the 'physical' world of such a device.

Like microelectronic devices, the scaling toward smaller dimensions is a keyfactor for the development and evolution of MEMS. In fact, the reduction of the scale size for microelectromechanical structures means denser (more structures on the same chip or substrate area) and faster (higher working frequency) devices, but it also represents the enabling step to a new set of phenomena, effects and potential novel applications. In particular, MEMS operate at a scale, the microscale, where the governing physical phenomena are the same as that of the 'macroworld', but in many cases they intervene with different weight with respect to the more common 'macroscale'.

A further 'shrinking' of both electronic and electromechanical devices will enter them into new dimensional scales, the 'mesoscale' and the 'nanoscale'. Some quantum-mechanical effects become observable and therefore the device models must take this into account. In the case of electronic devices, some of these quantum-mechanical effects may represent fundamental limits for further, future MOSFET scaling, while other effects provide the basic principles for new generations of electronic devices. In the case of electromechanical systems, this scaling step will provide higher sensitivities to alteration of the system's physical properties, plus suitable instruments for the study of the new phenomena that characterize such a mesoscale and which represent the key for the comprehension of phenomena at the nanoscale (atomic or molecular scale).

The characteristic feature size of a CMOS device is its channel length L, which is used as a performance index that identifies a given technology. In fact, the value of L for a CMOS technology identifies many important parameters, like the current-carrying capability, the power dissipation, switching speed and delay, etc., and therefore, a unique parameter, the channel length L can be assumed as a paradigm for the analysis of the scaling effects on CMOS technologies. In the last two decades, the scaling trend in CMOS technology has obeyed *Moore's law*. However, today's commercial CMOS technologies are characterized by channel lengths of the order of hundreds of nanometres or even smaller. In this particular size regime, some quantum-mechanical effects become observable (Taur, 2002; Frank, 2002) and degrade the device performance; thus, a value of 100 nm for the channel length L could be viewed as a limit for the validity of Moore's law and the scaling of CMOS technologies. On the other hand, the same challenging phenomena may represent an opportunity and the base for the development of novel classes of electronic devices, the future building blocks of electronic circuits (Goldhber-Gordon *et al.*, 1997; Roukes *et al.*, 2001; Lent *et al.*, 1993). Of course, this transition is not abrupt and many 'intermediate' solutions to 'migrate' from traditional CMOS technologies to future nanoelectronic devices have been adopted or are currently under development (Geppert, 2002).

The field of microsystems technology 'misses' an equivalent of the CMOS transistor and a characteristic parameter like the channel length

L. In fact, the variety of possible devices and functions implemented is very wide and it is quite impossible to find a common 'building block' and a unique parameter summarizing a large spectrum of technological, functional and scaling issues for all of the possible devices. It could be possible to classify MEMS by the function they implement, or the technology used, or both and to find, for each class a set of characteristic parameters which describes the most important features for that class. However, the lack of an equivalent of the CMOS channel length L in microsystems technology makes a scaling analysis more difficult from a general, system-level point of view.

A model for the analysis of scaling, both from a general and an application-specific point of view, will be proposed here as a paradigm for the scaling of microsystems.

Instead of a basic, simple building block, the proposed approach is based on the analysis of a complex system, made of several devices with different functions which are interconnected and cooperate in order to perform well-defined tasks. The analysis of scaling for such a complex system allows evaluating and taking into account both the effects of scaling on the single subsystems, and the relative influences each scaled subsystem has with each other. Moreover, this approach allows analysing scaling both from a general, system-level point of view, and from a more specific, device and application-level point of view. Therefore, the proposed idea could be a 'backbone' for the development of microsystem analysis and design methodology based on scaling laws.

The ideal candidate for the role of a scaling paradigm for microsystems is an autonomous microrobot of the type discussed previously in this chapter. In fact, an autonomous microrobot is a MEMS device which integrates almost all of the functionalities that can be implemented in a given technology or set of compatible technologies: sensing, actuation, control, signal processing and communication electronics and power supply. Hence, the analysis of an autonomous microrobot allow addressing many general design, technological and functional issues; moreover, the analysis can be easily specialized for a given device/application by taking into account only a subsystem; this process can then be extended down to the level of a single, simple device (sensor, actuator or circuit).

Furthermore, a robot is a system that is intrinsically multidisciplinary, in the sense that its design, realization and operation require multiple competences and the interaction between them. Therefore, the choice to consider a robot as reference system will allow analysing the scaling of multiple objects and from many points of view.

Many challenges are related to the scaling of a robot and they may be grouped into three categories: applications, methods and technologies, and physical principles. The reason why these three aspects of scaling are listed in an apparent 'reversed order' will be clarified below.

Despite the fact that the problem of scaling a complete and functional robot to the micrometre scale is an interesting and fascinating challenge in itself, from an engineering point of view it makes sense to think about the applications of such a miniaturized robot.

Once applications, actual or potential, have been envisioned, one should think about the methods and technologies that can be used to fabricate the microrobot or the single subsystems that comprise it.

At the microscale, the physical phenomena are the same as that of the macroscale, because the scale is still too large to observe quantum effects, although the relative weight of the phenomena may change. Therefore, some approximations that are valid at the macroscale may not be useful or accurate enough at the microscale; thus, proper model 'reviews' are required. Furthermore, due to this change of the relative influence of the physical phenomena, effectiveness of sensing and actuation means may change at the microscale to a degree that some methods gain in terms of sensitivity, while others may become impractical. So, the problem questioned here is relative to the degree of scaling that can be conveniently applied to a given sensing or actuation method or structure. An analogous problem exists for the energy source.

A robot having micrometre sizes and operating at the microscale is essentially a potential set of tools to perform many tasks at the same scale or even smaller. This is the basic concept, for example, of the 'microfactory': small machines used to fabricate and assemble smaller ones, in a chain that can be ideally extrapolated to smaller dimensions. Microrobots equipped with SPM or AFM probes can perform measurements, manipulations and assembly at the molecular or atomic scale. 'Microsurgery' is another interesting application field where microrobotic devices may find a wide range of uses: mini-invasive and localized surgery, targeted drug delivery, diagnostics and analysis. Inspection and maintenance, 'microsurveillance' systems and distributed measurements are other fields where microrobots can be effectively used.

The potential suitability of Si-MEMS-based microrobots for being batch-fabricated will provide the availability of large numbers of devices. This characteristic offers access to another range of applications where the co-operative behavior of a large number of identical systems, distributed control strategies, redundancy and population-like dynamics

Figure 7.3 Illustration of a 'population' of microrobots

can be explored and exploited. An example of a 'population' of micro-robots is shown in Figure 7.3.

The set of actual and potential applications mentioned above contributes to reinforce the already interesting and fascinating challenge of miniaturizing a robot.

The technologies of interest, in this work, for the fabrication of microsystems are the same used in microelectronics and are compatible with them, in particular with CMOS technology. In order to realize the suspended, moveable mechanical parts or the complex three-dimensional shapes required by MEMS, particular processing steps need to be added to the planar microelectronic technology. These processes are mainly aimed at releasing the mechanical parts from the substrate, to achieve the three-dimensional or quasi-three-dimensional geometries and, in some cases, to introduce and process some exotic materials in a conventional technology (for example, polymers or magnetic materials). The additional processing steps required to produce MEMS are usually termed 'micromachining' and are typically based on etching processes. Micromachining techniques are largely addressed in the literature and detailed descriptions of the technologies and related issues may be found in Kovaks (1998), Madou (2002) and Gad-El-Hak (2001).

Standard microelectronics technologies are planar processes and therefore they have the ability to define and pattern only two-dimensional geometries. Micromachining techniques allow realizing suspended structures with a certain control in the third dimension. However, in certain complex applications, like an autonomous micro-robot, more complex three-dimensional structures may be required; in such cases, some kind of assembly, manipulation or shaping strategy must be applied in order to achieve the required complexity. Examples of such approaches are given in Yeh and Pister (2000) and Eberfors *et al.* (1999).

REFERENCES

T. Ebefors, J.U. Mattsson, E. Mattsson, E. Kälvesten, G. Stemme (1999). A walking silicon micro-robot, in *Proceedings of the 10th Conference on Solid-State Sensors and Actuators (Transducers '99)*, Sendai, Japan, June 7–10,pp. 1202–1205.

D.J. Frank (2002). Power-constrained CMOS scaling limits, *IBM J. Res. Devel.*, **46**, 235–244.

M. Gad-El-Hak (2001). *The MEMS hand book*, CRC Press LLC, Boca Raton, FL, USA.

L. Geppert (2002). The amazing vanishing transistor act, *IEEE Spectrum*, **39**(10), 28–33.

D. Goldhaber-Gordon, M.S. Montemerlo, J.C. Love, G.J. Opiteck and J.C. Ellenbogen (1997). Overview of nanoelectronic devices, *IEEE Proc.*, **85**, 521–540.

G.T.A. Kovaks (1998). *Micromachined Transducers Sourcebook*, McGraw-Hill, New York, NY, USA.

C.S. Lent, P.D. Tougaw, W. Porod and G.H. Bernstein (1993). Quantum Cellular Automata, *Nanotechnology*, **4**, 49–57.

M.J. Madou (2002). *Fundamentals of Microfabrication: the Science of Miniaturization*, 2nd Edition, M. Gad-el-Hak (Ed.), CRC Press, Boca Raton, FL, USA.

M. Roukes (2001). Plenty of room, indeed, Scientific American Magazine, September 9, pp. 48–57.

Y. Taur (2002). CMOS design near the limit of scaling, *IBM J. Res. Devel.*, **46**, 213–222.

R. Yeh and K. Pister (2000). Design of low-power articulated microrobots, in *Proceedings of International Conference on Robotics and Automation Workshop on Mobile Micro-Robots*, San Francisco, CA, USA, April 23–28, pp. 21–28.

8

Moving towards the Nanoscale

8.1 SEMICONDUCTOR-BASED NANO-ELECTROMECHANICAL SYSTEMS

It has been anticipated in Chapter 1 that when the device feature size shrinks down to the sub-micrometre range some quantum-mechanical effects may become observable, while classical physics laws, and then scaling laws as they have been intended throughout this work, break. This particular size regime, the mesoscale, and the smaller one, the nanoscale, are beyond the scope of this work; however, advances in some fabrication techniques have allowed realizing feature sizes in the nanometre range, like *electron-beam lithography*. Moreover, developments of MEMS technologies and devices have provided a platform for the fabrication of 'microtools' which, in turn, could be used for the fabrication of smaller devices, such as scanning-probe microscophy.

Since MEMS technologies and devices have the potential to be employed in the realization of structures operating at the meso- and nanoscale levels, an overview of semiconductor-based Nano-Electro-Mechanical Systems, or NEMS, which have been reported in the literature, aimed at the study and exploitation of characteristic phenomena and effects at the meso- and nanoscale levels, is presented in this chapter.

All of the devices taken into account in this overview, apart from the differences in application fields and operating principles, are characterized by very simple structures. In fact, in comparison to MEMS, nano

Scaling Issues and Design of MEMS   S. Baglio, S. Castorina and N. Savalli
© 2007 John Wiley & Sons, Ltd

electromechanical devices are mainly based on simple structures like bridges, cantilevers and plates; they usually have simple geometries and limited degrees of freedom. However, the structural simplicity of such nanodevices is an indicator of the early stage of evolution of this field, and it does not mean that such devices are useless or limited to a demonstrative use. Despite their simple structures, some of the devices presented here have been used to successfully investigate novel and interesting phenomena at the meso- and nanoscales, and they represent a further step toward novel application fields.

8.2 NANOFABRICATION FACILITIES

In the realization of nanodevices, almost the same process steps used in semiconductors microtechnology are employed: oxidation and deposition processes, wet and dry etching processes, evaporation, sputtering, etc. What is essentially different in such a fabrication approach is the method to define the nanometric feature sizes on the device substrate. The electron-beam lithography and the *Atomic Force Microscophy/Scanning Probe Microscophy* (AFM/SPM)-based lithographic methods are powerful tools which allow achieving the nanometric range resolution required for the fabrication of nanodevices. These nanofabrication facilities will be briefly introduced in the following. Information on electron-beam lithography and AFM/SPM patterning are reported in Bernstein *et al.* (2002).

In electron-beam lithography, an electron beam is used to selectively change the properties of a sensitive film deposited on the surface of the substrate. The issues related to the high resolution of electron-beam lithography are the same as those of high-resolution scanning electron microscopy (SEM), which is a mature field; therefore, electron-beam lithography apparatus are commonly based on modified SEM.

The operation of electron-beam lithography is as follows: the substrate is covered with a thin layer of an electron-sensitive material, for example PMMA and then it is exposed to the electron beam; the exposed PMMA changes its solubility towards certain chemicals; a suitable metallization can be deposited in developed PMMA areas, which can serve as a structural layer, or as a mask against successive etching steps.

Higher resolution, close to the atomic scale, can be achieve with AFM/SPM-based techniques. The very sharp tip of a scanning probe microscope, with suitable controls, can be used to selectively remove hydrogen atoms from the hydrogen-passivated surface of a silicon substrate, thus achieving atomic resolution.

However, despite the higher resolution achieved with SPM, many challenges remain before it can be applied to the large-scale fabrication of NEMS. First of all, the 'fragility' of the produced patterns, which makes it difficult to retain the high resolution throughout the fabrication of three-dimensional structures. Moreover, such a technique has an extremely low throughput.

In AFM-based lithography, the microscope tip can be either used as a 'scriber' or to locally activate silicon or metal oxidation. The resolution achievable with AFM is lower than SPM, since it is limited by the shape and size of the tip, especially in the scribing mode, where the tip is used to scratch the substrate surface, and then it is subject to wear. In order to limit wear, the surface is usually covered with a soft material, like PMMA, which can be displaced by the tip.

8.3   OVERVIEW OF NANOSENSORS

Among the nano-electromechanical systems reported in the literature, some show potential to be used as nanosensors. They can be distinguished on the basis of the measurand, the operating principle, the detection method, the structure and so on. In this overview, a fundamental classification in SPMbased nanosensors and standalone nanosensors has been followed. To the first class belong all of the applications in which the measurement of a given physical property at the nanoscale is performed with a suitably functionalized scanning probe microscope tip, whereas the stand-alone sensors are all those devices that embed all of the functional parts in their structures and require only excitation and the detection equipment to be provided.

It will be highlighted here how scanning probe microscopes are flexible tools which can be used in fabrication, analysis and characterization of nanometre-scale devices.

In the application examples reported here, atomic force microscopes and scanning tunnelling microscopes are used as force/displacement sensors, thermal and magnetic imagers, and biological mass sensors. This latter application does not 'roughly' use AFM or STM equipment as in the other examples, but the detection method and equipment typical of scanning probe microscopy are 'grabbed' to detect the motion of micronano cantilevers functionalized with immunospecific biological detectors.

8.3.1   Use of AFM for materials and nanodevices characterization

The atomic force microscope can be used to characterize materials properties down to the nanoscale. An example of such investigations is presented in Namazu *et al.* (2000), where a characterization method to reveal the specimen size effect on the Young's modulus and bending strength of single-crystal silicon down to the nanoscale is presented. In this work, AFM is used to carry out bending tests on nanometre-scale silicon fixed beams. Moreover, AFM has also been used for the fabrication of the nanobeams through the field-enhanced anodization method.

The test was executed by first measuring the beam dimensions through the microscope and then by executing the bending test through a calibrated microscope cantilever with a diamond tip. The Young's modulus and bending strength of silicon beams were finally estimated by equations based on the assumption that the beams follow the linear elastic theory of an isotropic material.

Experimental results have shown that the Young's modulus of silicon has no specimen size effect in the range from a nano- to a millimetre scale. This is attributed to the uniform microstructure of silicon, regardless of the specimen size in which the shape of the interatomic energy curve between the silicon atoms does not change. It has also been observed that the bending strength increases with a decrease of the specimen size. Further mechanical properties characterization experiments for nanostructures can be found in Sundararajan and Bhushan (2002).

8.3.2   Scanning thermal microscopy (SThM)

The investigation of materials thermal properties at the nanometre scale is of high interest, both in the scientific and industrial fields. The mapping of thermal properties on a surface can be executed with a modified scanning tunnelling microscope, equipped with a suitable tip. This type of scanning thermal microscope is reported in Lee *et al.* (2002). In this work, a batch-fabrication method for the realization of a thermocouple on an insulating tip is proposed. This junction on the tip is used to induce local heating and then to measure local temperature variations. Further application of such an SThM tool in probe-based data storage systems is currently under investigation.

8.3.3   Scanning Hall probe microscopy

Another example of a modified scanning probe microscope tip for sensing application employs Scanning Hall Probe Microscopy (SHPM), as reported in Chang *et al.* (1992) for the measurement of magnetic fields and properties at the nanoscale. Many potential applications are possible, such as the study of surface magnetic memory devices, or the magnetic field profiles due to vortices in superconducting films, networks, or crystals, spatially varying current flow in patterned microstructures, or domain structures near surfaces in magnetic systems. This tool combines the high-field sensitivity of submicron heterostructures, with the high-precision positioning offered by the SPM technique. Field sensitivities of $0.36\,\text{GHz}^{-1/2}$ at 4.2 K, with a spatial resolution of $0.35\,\mu\text{m}$, have been demonstrated.

8.3.4   Mechanical resonant immunospecific biological detector

The application of micro- and nano-electromechanical systems to chemical and biological sensing has received great interest in the scientific and industrial communities. For example, micro- and nano-mechanical oscillators have been proposed as resonant chemical and biological sensors. These approaches are usually based on a mechanical structure, the oscillator, coupled with a very sensitive displacement detection scheme. Among the proposed approaches, several scanning force microscopy techniques have been successfully applied.

As an example, the system described in Ilic *et al.* (2000) is a resonant mass detection biological sensor, which is composed of an array of silicon nitride cantilevers as resonant structures and an optical deflection system for the measurement of the out-of-plane vibrational mode. In particular, in this application a laser beam excites the vibrational mode of the cantilever due to thermomechanical noise. The reflected light is detected by a photodiode which serves as a position-sensitive detector.

In order to determine the mass bound to each of the cantilevers in the array, frequency spectra were taken before and after binding of the cells to the antibodies. When a binding event occurred, the additional cell mass loading caused a shift in the resonant frequency of the micromechanical oscillator. The mass sensitivity of the cantilevers depends on the mechanical quality factor, Q, of the device, which can be determined from the bandwidth at resonance. Mass sensitivities as low as 5.1 Hz/pg have been reported and the minimum detectable mass of the cantilevers in vacuum is 14.7×10^{-15}g.

8.3.5 Micromechanical sensor for differential detection of nanoscale motions

Another approach for biological mass detection is presented in Savran *et al.* (2002). Here, two identical cantilevers have been realized, but only one of them has been functionalized with immunospecific receptors. By using a differential detection scheme, the difference between the deflection signals of the two cantilevers can be read, while still minimizing the impact of temperature. An interdigited diffraction grating has been realized between the two cantilevers to allow direct optical reading of the displacement through interferometric techniques.

The differential measurement proposed in this work is much less affected by temperature changes when compared with the absolute one. Since both cantilevers have the same thermal response, detecting the differential bending can easily reduce the effect of temperature fluctuations and provide a more reliable detection of specific bending caused by a biochemical/chemical reaction.

8.3.6 Nanomagnetic sensors

We will provide here a first example of a stand-alone nanosensor. In particular, this employs a small Hall probe for the characterization of magnetic surfaces, which has been demonstrated in a scanning probe arrangement in Monzon *et al.* (1999). In this work, an approach exploiting the local Hall effect (LHE) is proposed, which makes use of standard and electron-beam lithography processes. The device is based on a small ferromagnet, deposited via a 'lift-off' process, over a semiconducting cross-junction, with an edge positioned in the centre of the cross-junction. A small AC bias current is applied to the cross-junction while an external in-plane magnetic field is varied. Fringe fields from the edge of the ferromagnet induce an AC Hall voltage that is directly proportional to the magnitude and direction of the magnetization. A series of measurements has been made on individual nanomagnets having a range of aspect ratios and widths, in a swept magnetic field, H. The shape of the hysteresis loops, and the coercivity, Hc, both vary greatly with aspect ratio and width.

8.3.7   Nano-wire piezoresistors

The piezoresistive effect is a widely exploited sensing principle in MEMS and has been used, for example, in pressure sensors and accelerometers. Moreover, piezoresistive sensing is one of the detection methods used in the AFM cantilever, together with optical detection. In particular, due to its lower sensitivity with respect to optical detection, piezoresistive sensing is used in those applications where optical detection is difficult to apply, for example, when large array of devices are used, as in data storage systems, biochemical and DNA mass sensing.

A scaling analysis performed on an ideal piezoresistor with a rectangular shape revealed that in order to increase sensitivity it is necessary to reduce the thickness and mass of the piezoresistor (and cantilever), while still keeping its length constant. In particular, in order to achieve force resolution below fN, which is necessary to detect masses of molecules and ions, the thickness and mass of the piezoresistor and the cantilever must be reduced to the nanoscale.

In Toriyama *et al.* (2002), the pizoresistance in *p*-type single crystal silicon nanowires (Si nanowires) has been studied in order to verify the performance of a nanometric piezoresistor. It has been found in this case that the longitudinal piezoresistance coefficient increased with a decrease in the cross-sectional area, while the transverse piezoresistance coefficient decreased with a increase in the aspect ratio of the cross-sectional area, and has a very weak dependence on its shape and size.

Experiments have shown that the relative resistance changes in nanowire piezoresistors vary linearly with the applied stress at room temperature.

8.3.8   Nanometre-scale mechanical resonators

Nanometre-scale mechanical resonators can be fabricated from single-crystal silicon substrates by means of electron-beam lithography and silicon micromachining techniques, as shown in Cleland and Roukes (1996), where the fabrication and characterization of nanometre-scale mechanical resonators, which can be applied to several nanosensing applications, are reported.

Beyond very high resonance frequencies, nanometre-scale mechanical resonators are small enough so that it will be extremely improbable that their structures contain crystalline defects and therefore these nanometre-scale mechanical resonators also exhibit very high quality factors.

Such devices can be used as particle and energy sensors, thanks to their small mass and size, high operating frequency and sensitivity to external conditions. Furthermore, they could exhibit some macroscopic quantum effects at very low temperatures.

In Cleland and Roukes (1996), several suspended beams have been fabricated and characterized as mechanical resonators at a very low temperature (4.2 K), by means of a magneto-motive measurement method. Resonance frequencies up to 120 MHz and quality factors up to 10^4 have been reported.

An example of the application of a nanomechanical resonator as a 'nano-displacement' sensor is reported in Ekinci et al., (2002). The proposed sensor makes use of a magneto-motive detection method. In fact, the magnetic, electrostatic and piezoresistive transduction methods, which have been successfully applied at the microscale, may become insensitive at the submicron scale. Moreover, unavoidable stray couplings may make unusable the conventional detection schemes at the nanoscale. Instead, magneto-motive detection scales well in the nanoscale and offers direct electronic coupling to the resonator displacement. The response of a sample sizing, $15 \mu m \times 500 nm \times 350 nm$, has been reported; the device has a resonance frequency higher than 25 MHz and a quality factor higher than 10^4 at 20 K.

Measurements on doped sample have also been performed in order to investigate dissipation mechanisms due to surface adsorbates and defects.

8.3.9   Electric charge mechanical nanosensor

A further example of a mechanical resonator has been reported in Cleland and Roukes (1998), where it is applied as an electric charge sensor. The nanomechanical electrometer consists of a compliant mechanical element, electrodes and a displacement detector. The presence of a small charge on the electrode affects the motion of the mechanical resonator and these changes are revealed by the displacement detector, which is based on the magneto-motive method. Frequency modulation detection has been employed as a measurement technique, in order to obtain enhanced sensitivity and large measurement bandwidth.

These devices have been characterized as charge detectors, under vacuum conditions at 4.2 K. Results have shown a resonance frequency shift from 2.6115 to 2.6170 MHz, due to a total charge variation of 4×10^4 e.

8.4 CONCLUDING REMARKS

Some examples of nanodevices and applications have been provided in this chapter. The treatise does not pretend to be exhaustive, mainly because the field of nanotechnology evolves at a very high rate and new devices and applications may arise everyday. However, the examples presented here can provide an insight into the silicon-related nanotechnologies, devices and applications, to show some of the potentials of this field and also to provide an idea of the current state of the art in the field.

REFERENCES

G. Bernstein, H.V. Goodson and G. Snider (2002). Fabrication Technologies for Nanoelectromechanical Systems, in *The MEMS Handbook*, M. Gad-el-Hak (Ed.), CRC Press, Boca Raton, FL, USA.

A.M. Chang, H.D. Hallen, L. Harriott, H.F. Hess, H.L. Kao, J. Kwo, R.E. Miller, R. Wolfe, J. Van der Ziel and T.Y. Chang (1992). Scanning Hall Probe Microscopy, *Appl. Phys. Lett.*, 61, 1974–1976.

A.N. Cleland and M.L. Roukes (1996). Fabrication of high frequency nanometer scale mechanical resonators from bulk Si crystals, *Appl. Phys. Lett.*, 69, 2653.

A.N. Cleland and M.L. Roukes (1998). A nanometre scale mechanical electrometer, *Nature*, 392, 160.

K.L. Ekinci, Y.T. Yang, X.M.H. Huang and M.L. Roukes (2002). Balanced electronic detection of displacement in nanoelectromechanical systems, *Appl. Phys. Lett.*, 81, 2253.

B. Ilic, D. Czaplewski, H.G. Craighead, P. Neuzil, C. Campagnolo and C. Batt (2000). Mechanical Resonant Immunospecific Biological Detector, *Appl. Phys. Lett.*, 77, 450.

D.W. Lee, T. Ono and M. Esashi (2002). Fabrication of thermal microprobes with a sub-100 nm metal-to-metal junction, *Nanotechnology*, 13, 29–32.

F.G. Monzon, D.S. Patterson and M.L. Roukes (1999). Characterization of individual nanomagnets by the local Hall effect, *J. Magnet. Magn. Mater.*, 195, 19–25.

T. Namazu, Y. Isono and T. Tanaka (2000). Evaluation of Size Effect on Mechanical Properties of Single Crystal Silicon by Nanoscale Bending Test Using AFM, *J. Microelectromech. Syst.*, 9, 450–459.

C.A. Savran A.W. Sparks, J. Sihler, Jian Li, Wan-Chen Wu, D.E. Berlin, T.P. Burg, J. Fritz and M.A. Schmidt (2002). Fabrication and Characterization of a Micromechanical Sensor for Differential Detection of Nanoscale Motions, *J. Microelecmech. Syst.*, 11, 703–708.

S. Sundararajan and B. Bhushan (2002). Development of AFM based techniques to measure mechanical properties of nanoscale structures, *Sensors Actuators A: Phys.*, 101, 338–351.

T. Toriyama, Y. Tanimoto and S. Sugiyama (2002). Single Crystal Silicon Nano-Wire Piezoresistors for Mechanical Sensors, *J. Microelectromech. Syst.*, 11, 605–611.

9

Examples of Scaling Effects Analysis – DIEES-MEMSLAB

9.1 INTRODUCTION

This chapter gives a few guidelines to make practice with MEMS and scaling effects. It starts by dealing with a simple reference mechanical configuration, a bylayer cantilever beam and then reports a tutorial of a software tool, conceived to make simulations on micromachined plates (membranes) and interdigitated devices.

In particular, in the first part of this chapter, the theory of scaling applied to the simple reference MEMS structure allows us to verify with concrete numerical examples the effects of scaling dimensions on the structure's performance as a sensor or an actuator.

Simplifying assumptions are made considering two layers with the same dimensions. Linear dimensions scale with factors of 10 and 100 from an initial choice and the effects of such scaling to the mechanical, electrical and thermal characteristics are investigated and compared. Moreover, the consequences of reducing the structure's size on its behaviour as a sensor (force, displacement, mass, acceleration) or an actuator (electrostatic and thermal) will also be examined.

On the other hand, the development of the proposed simulator aims to provide graduating students in studies of an efficient instrument to better understand the mechanical modelling of more complex microdevices, and to derive considerations about scaling effects. To this purpose, it

Scaling Issues and Design of MEMS   S. Baglio, S. Castorina and N. Savalli
© 2007 John Wiley & Sons, Ltd

has been written with the target of being an 'open source-code', to be optimized and extended with the help of other students worldwide.

Although it is not easy to take into account the modelling issues of every microdevice actually proposed as a sensor or actuator, it makes sense the proposal to develop a relatively simple code for investigating the main mechanical configurations for microelectromechanical systems. In particular, the intermediate target has been that of including software procedures used to study static and dynamic behaviour of the following reference devices:

(1) *Plates*. Both thin plates (say membranes) or bulky devices are just taken into account, with particular attention in considering the deviation from ideal (homogeneous) behaviour for structures composed of a stack of different materials. A reference mechanical configuration has been assumed to be made by a device composed of a central part (suspended plate) anchored to the substrate by means of four symmetric 'flexures'. Implementation of the equivalent section method improves in this sense the accuracy of results inherent to the estimation of mechanical parameters (such as elastic constants). To this purpose, most of the work has been made for developing models on structures operated as inertial sensors (mainly accelerometers). The effects of scaling geometrical features can be appreciated by running simulations. The code to study masssensors, such as those depicted in Chapter 5, will be added soon.

(2) *Interdigitated Structures*. In this case, a reference structure made of a central plate with appendices, being the rotor fingers and four symmetric 'flexures', has been considered. Capacitive sensors and actuators can be hence both analysed.

It is worth while to observe that the proposed software tool aims only to provide an interactive instrument to professors of basic courses on microelectromechanical systems. The large number of ideal hypotheses is still very high, since high-order dynamic effects, or non-linearities are not included. In addition, real operating conditions of the conceived structures, such as those coming from etching procedures, were not, neither can be easily, taken into account (think, for example, of the 'curling effects' in interdigitated devices, which strongly affect the final active area extension between interdigitated fingers).

Furthermore, since basic courses on MEMS often start from analytical modelling approaches, the proposed simulator fits well with the needs of beginners in this field, and being an open source-code can be easily adapted to their own needs.

Finally, these authors strongly believe that the proposal of a synergic development of MEMSLAB, with student readers, could be the right way to provide them a significant support in the initial understanding of such a huge 'microworld'.

9.2 EXAMPLES OF SCALING CANTILEVER BEAM DEVICES

In order to simplify the following analysis, a simplified bilayer structure has been considered with given proportions between its dimensions. The structure taken into account here is shown in Figure 9.1.

The cantilever beam's dimensions will be considered to be as follows:

$$L = 10W; W = 10t$$

The scaling effects on such kind of structures will be examined on three different dimensional scales, as reported in Table 9.1.

Some properties of some of the materials commonly available in most common integration technologies are reported in Table 9.2.

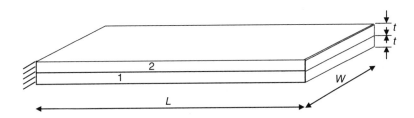

Figure 9.1   Structure of a bilayer cantilever beam

Table 9.1   Dimensions of the structures

Case	L (m)	W (m)	t (m)
i	10^{-2}	10^{-3}	10^{-4}
ii	10^{-3}	10^{-4}	10^{-5}
iii	10^{-4}	10^{-5}	10^{-6}

Table 9.2 Properties of the materials

Material	Density, $\rho(Kg/m^3)$	Young's modulus, Y (Pa)	CTE, $\alpha(K^{-1})$	Thermal conductivity, $k(Wm^{-1}K^{-1})$	Heat capacity, $c(J Kg^{-1}K^{-1})$
SiO$_2$	2.5×10^3	7.3×10^{10}	4×10^{-7}	1.1	—
Si/PolySi	2.3×10^3	1.9×10^{11}	2.5×10^{-6}	160	700
Metal	2.7×10^3	7.0×10^{10}	2.3×10^{-5}	240	900

The elastic constant, K_z, for the structure depicted in Figure 9.1 is given by:

$$K_z = \frac{F_z}{\delta_z} = \frac{12E_n I_{ny}}{L^3} \tag{9.1}$$

Where E_n is the highest Young's modulus between the layers and I_{ny} is the moment of inertia of the cantilever's normalized, or equivalent, cross-section, which is given by:

$$I_{ny} = \sum \left[\frac{W_i t_i^3}{12} + S_i \left(h_n - h_i \right)^2 \right] \tag{9.2}$$

where $W_i = WE_i/E_n$ is the normalized width of the ith layer.

If one assumes that the cantilever is made of a first layer of polysilicon and a second layer of metal, with the structure depicted in Figure 9.1 and the dimensions reported in Table 9.1, this results in $E_n = 1.9 \times 10^{11}$ Pa

In equation (9.2), S_i is the area of the normalized cross-section, that is, $S_i = W_i t_i$, h_i is the height of the median axis of the ith layer and h_n is the height of the cantilever's neutral axis, which is defined as:

$$h_n = \frac{\sum S_i h_i}{\sum S_i} \tag{9.3}$$

For the three cases considered above, we obtain the data presented in Table 9.3.

The data given in Table 9.3 show that the cantilever's elastic constant, K_z, linearly scales with its dimensions. A comparison of the force required to displace the cantilever tip by a given amount, say 10% of the cantilever length, gives the data presented in Table 9.4.

From the data given in Table 9.4 it is clear that the mass, M, scales as L^3, while the elastic constant K_z scales as L; therefore, if the dimensions

Table 9.3   Results obtained from the normalized cross-section method

Case	Normalized width (m)		Normalized cross-section (m^2)		Neutral axis, h_n (m)	Moment of inertia, I_{ny} (m^4)	Elastic constant, K_z (N/m)
	W_1	W_2	S_1	S_2			
i	10^{-3}	3.68×10^{-4}	10^{-7}	3.68×10^{-8}	7.69×10^{-5}	7.21×10^{-16}	1.64×10^3
ii	10^{-4}	3.68×10^{-5}	10^{-9}	3.68×10^{-10}	7.69×10^{-6}	7.21×10^{-20}	1.64×10^2
iii	10^{-5}	3.68×10^{-6}	10^{-11}	3.68×10^{-12}	7.69×10^{-7}	7.21×10^{-24}	16.4

Table 9.4   Comparison of force, mass and energy between the three cantilever structures

Case	Displacement, δ (m)	Force, F_z (N)	Energy, $W_m/$(J)	Volume, V (m^3)	Energy density, w (J/m^3)	Mass, M_0 (kg)
i	10^{-3}	1.64	1.64×10^{-3}	2×10^{-9}	8.2×10^5	5×10^{-6}
ii	10^{-4}	1.64×10^{-2}	1.64×10^{-6}	2×10^{-12}	8.2×10^5	5×10^{-9}
iii	10^{-5}	1.64×10^{-4}	1.64×10^{-9}	2×10^{-15}	8.2×10^5	5×10^{-12}

are scaled down by a factor of 10, a structure 1000 times lighter and only 10 times less stiff is achieved – thus, the overall mechanical robustness of the structure is improved. It can also be observed that the force scales as L^2, the energy scales as the volume while the energy density has obviously been kept constant.

The behaviour of the cantilever as a resonant device will be taken into account in the following. The results obtained for the three cases considered above are presented in Table 9.5.

It can be seen that the resonance frequency scales as $1/L$.

Equation (1.5) in Chapter 1 gives the dependence of the cantilever's resonance frequency from the variations, Δm, of the cantilever mass M_0:

$$\omega = \frac{\omega_0}{\sqrt{1 + \frac{\Delta m}{M_0}}} \qquad (9.4)$$

Table 9.5   Resonance frequency data

Case	ω_0 (rad/s)
i	1.8×10^4
ii	1.8×10^5
iii	1.8×10^6

Table 9.6   Resonance frequency shifts due to mass changes

Case	Mass, M_0 (kg)	Resonance frequency, ω_0 (rad/s)	Mass change, Δm (kg)	Frequency shift, $\Delta\omega$ (rad/s)	Sensitivity, $\Delta\omega/\Delta m$ (rad/s/kg)
i	5.0×10^{-6}	1.81×10^4	5.0×10^{-9}	9.06	1.81×10^9
ii	5.0×10^{-9}	1.81×10^5	5.0×10^{-12}	90.06	1.81×10^{13}
iii	5.0×10^{-12}	1.81×10^6	5.0×10^{-15}	9.06×10^2	1.81×10^{17}

and therefore, the relative variation in frequency results in the following:

$$\frac{\Delta\omega}{\omega_0} = 1 - \frac{1}{\sqrt{1 + \frac{\Delta m}{M_0}}} \tag{9.5}$$

If the cantilever mass varies, for example, by an amount equal to one-thousandth of its original mass $(0.001\,M_0)$, the respective changes in frequency for the three cases amount to those shown in Table 9.6.

The values given in Table 9.6 indicate that dimensional scaling leads to a higher sensitivity to absolute mass variations in terms of resonance frequency shifts. Then, a scaled structure, i.e. to micrometre or submicrometre scale, exhibits a potential high performance as a small mass sensor. As already shown in Chapter 1, a scale of a factor 10 in dimensions, leads to a 10 000-fold increase in sensitivity to mass changes in terms of the resonance frequency shift.

In Chapter 1, it has also been shown that a cantilever exited with a time-varying force, say, for example, sinusoidal, may be used as an acceleration sensor. In the particular case of constant, or slowly varying accelerations, the sensitivity in terms of displacement of the cantilever's average position versus the acceleration is expressed by equation ((1.3)) in Chapter 1, which in absolute values is as follows:

$$Sa = \frac{dx}{da} = \frac{M_0}{K_z} \tag{9.6}$$

For the three example considered here, we obtain the values given in Table 9.7.

A scaling factor of 10 in dimensions leads to a 100-fold reduction of sensitivity to acceleration, as already predicted in Chapter 1; therefore, scaling of mechanical structures leads to less susceptible devices against abrupt and/or strong accelerations or, in other words, scaled devices are more suited for the measurement of higher accelerations.

Table 9.7 A resonant cantilever
as an accelerometer

Case	Sa (m/m/s^2)
i	3.04×10^{-9}
ii	3.04×10^{-11}
iii	3.04×10^{-13}

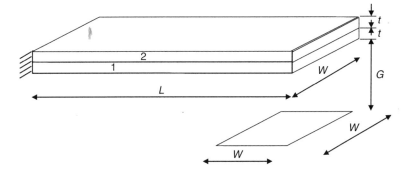

Figure 9.2 A cantilever as an electrostatic actuator

The use of a simple cantilever beam structure of the type described above as an electrostatic actuator will be examined here. Consider an electrostatic actuator where the two electrodes are realized by means of the cantilever beam and a fixed electrode, sized $W \times W$, placed under the tip of the cantilever, at a distance, G, as shown in Figure 9.2.

For this structure, the electrostatic energy stored in the air gap, as stated in equation (1.17) of Chapter 1, in terms of the electrostatic field in the gap, is:

$$W_E = \frac{1}{2}CV^2 = \frac{1}{2}\varepsilon_0 \frac{W^2}{G}V^2 = \frac{1}{2}\varepsilon_0 W^2 G E^2 \qquad (9.7)$$

The electrostatic energy is 'upper-limited' by the breakdown field of the dielectric between the capacitor's electrodes and scales as its volume. The attractive force on the upper, mobile electrode is given by equation (1.19) in Chapter 1, which is:

$$F_E = -\frac{\partial W_E}{\partial z} = \frac{1}{2}\varepsilon_0 W^2 E^2 \qquad (9.8)$$

In order to achieve some numerical examples for the three cases taken into account in the previous discussion, a value for G must be

Table 9.8   A cantilever as an electrostatic actuator

Case	Gap, G(m)	Energy, W_E (J)	Energy density, E_D (J/m^3)	Force, F_E (N)	Pressure, P_E (N/m^2)
i	2×10^{-3}	7.965×10^{-10}	3.98×10^{-1}	3.98×10^{-7}	3.98×10^{-1}
ii	2×10^{-4}	7.965×10^{-13}	3.98×10^{-1}	3.98×10^{-9}	3.98×10^{-1}
iii	2×10^{-5}	7.965×10^{-16}	3.98×10^{-1}	3.98×10^{-11}	3.98×10^{-1}

determined. Since a displacement of 10 % of L has been taken as a reference, and since $W = L/10$ has been supposed, G needs to be greater than this value; therefore, it can be supposed here that $G = 2W$.

The results obtained for the three cases, if a maximum allowed electric field of 3×10^6 V/m is applied, are shown in Table 9.8.

The data given in Table 9.8 confirms the results previously reported in Chapter 1.

If the Paschen effect occurs in case (iii), the maximum electrostatic breakdown field can grow up to 3.0×10^8 V/m in air, thus leading to a stored electrostatic energy of 7.96×10^{-12} J and an electrostatic force of 3.98×10^{-7} N. Therefore, electrostatic energy, energy density, force and electrostatic pressure are four orders of magnitude higher if the Paschen effect occurs. This is clearly a kind of 'brutal estimation' but gives an example of how 'isometric' scaling laws do not apply in certain cases, and can lead to a better overall performance. In fact, in this case, the scaled microelectrostatic actuator can perform much better than which is predicted with 'bare' geometric scaling laws, thanks to the relevance gained by physical effects usually negligible on the macroscale.

The cantilever shown in Figure 9.2 can also be exploited as a thermal actuator. In fact, due to the differences in thermal expansion coefficients of the two layers, when subjected to a uniform temperature variation ΔT, the cantilever tip displaces by an amount δ and can then perform 'useful' work. The relation between the temperature change and the tip displacement has been introduced in Chapter 2, and is given as follows:

$$\delta = L^2 \frac{3(1+m)^2}{t\left[3(1+m)^2 + (1+mn)\left(m^2 + 1/mn\right)\right]}(\alpha_2 - \alpha_1)\Delta T \qquad (9.9)$$

where m, n and t are, respectively, $m = t_1/t_2$, $n = Y_1/Y_2$ and $t = t_1 + t_2$.

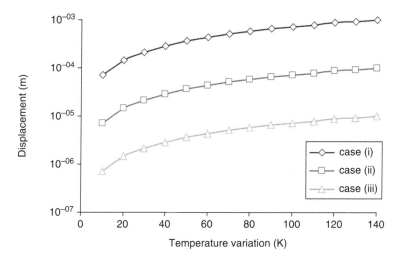

Figure 9.3 Cantilever tip displacement versus temperature variation in an application as a thermal actuator

By taking into account the three cases examined here, together with the materials properties reported in Table 9.2, the tip displacement as a function of temperature variation is reported in the plot of Figure 9.3.

At $\Delta T = 140\,K$, this corresponds to a tip displacement of approximately $10\,\%L$, which is a value assumed as a 'reference' throughout this analysis. In these latter conditions, the energy stored in the actuator amounts to the following:

$$W_T = \rho c V \Delta T = (\rho_1 c_1 + \rho_2 c_2) \, V \Delta T \qquad (9.10)$$

as shown in Table 9.9.

Table 9.9 reports the values of stored thermal energy for the three structure examples, for a temperature variation of 140 K, which corresponds to a tip displacement of $10\,\%L$. Moreover, the thermal energy

Table 9.9 Stored thermal energy and comparison with electrostatic energy

Case	Thermal Energy, W_T (J)	Electrostatic Energy, W_E (J)
i	5.66×10^{-1}	7.96×10^{-10}
ii	5.66×10^{-4}	7.96×10^{-13}
iii	5.66×10^{-7}	7.96×10^{-16}

is compared to the electrostatic energy required to achieve the same displacement in the application of an electrostatic actuator. These results show how such energy is much higher for thermal actuators, but this means that these actuators can potentially exert higher forces and/or perform much useful work. This is clearly an advantage of thermal actuators, especially on the microscale.

9.3   DIEES-MEMSLAB-TUTORIAL

9.3.1   Introduction

The simulator has been realized by using 'Mathworks Matlab 6.5' by graduating students afferent to the 'Measurement for the Automation and the Industrial Production' group, 'Microsystem' Division, at DIEES, University of Catania, Italy .

This has been conceived as a developing platform to be used for graduate and post-graduate students who just have a sufficient level of confidence with MEMSs issues. It is, in fact, thought of as a software ambient by which many aspects related to analytical modelling of MEMS and scaling issues for several of the most common device configurations can be analysed. For this reason, both actuators and sensors have been included among the examined examples; moreover, the spirit of such software support is prevalently that of stimulating other students, approaching the multidisciplinary world of MEMS, to continue developing the software in interaction with the book authors and students from other countries, hopefully worldwide. Although the accuracy of results coming from these simulations are quite far from those obtainable with finite-element analysis, a quick and easy approach to the most important issues in modelling and designing MEMS is proposed, with particular emphasis on the possibility of examining the effects of scaling dimensions.

Such tools allow us to make simple exercises and observe the effects of scaling for three mechanical topologies of microdevices, mainly consisting of:

- Cantilever beams: with devices anchored to one of their ends.
- Flexures: devices anchored to four symmetric supports at their corners (with two different shapes for supporting beams).

- Interdigitated devices: comb structures composed of a central part (plate), a set of fingers attached to the plate (rotor's fingers) and a set of fingers attached to the substrate (stator's fingers).

The above represent the first selection to be made through the main interface shown in Figure 9.4.

For the above listed types of devices, different actuation or sensing principle can be analysed, and in particular:

- The cantilever structure has been taken into account for examining bilayer thermal actuators, electrostatic actuators, accelerometers and resonant mass sensors.
- the flexure are considered for examining: accelerometers and mass sensors.
- the interdigitated devices are modelled for examining: electrostatic actuators and accelerometers with capacitive output.

Therefore, once the reference mechanical structure is selected, and a choice is made between sensor or actuator, a selection regarding the

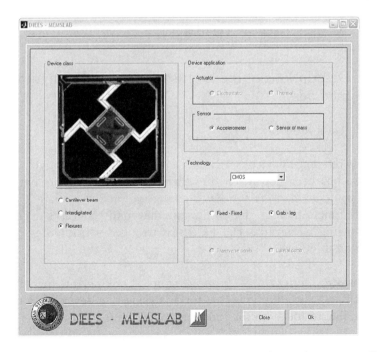

Figure 9.4   Main window of DIES-MEMSLAB. First, a selection between cantilever beams, flexures or interdigitated devices must be made

realization process and hence the number and physical parameters of the materials comprising the device, has to be addressed. The following list is applicable:

- Homogeneous – one single material composes the structures and a free choice for micromachining procedures is supposed to be imagined by the user.
- CMOS AMS 0.8 μm, including post-processing micromachining by means of wet TMAH procedures.
- An SOI-based process in which the wafer is supposed to be consequently DRIE-etched from both the back side and the front side.
- A CUSTOM process in which the number of layers, their thickneses, mechanical and electrical properties of the layers, and the micromachining procedures can be imagined by the user (a maximum of ten layers can be used).

A particular case is represented by the bilayer thermal actuator in which a simplified technology with two layers is considered.

Ulterior selections in the master window regard the interdigitated electrostatic device, which can be lateral or transverse and the shape of the supporting beams for mechanical devices, as flexures or interdigitated, that can be 'fixed–fixed' or 'crab-leg'.

9.3.2 Descriptions of the microstructures and analytical methods

9.3.2.1 Cantilever Beams

The first reference device is a cantilever beam. Such a relatively simple structure has been considered in other chapters of this present book. It is considered here as the basic configuration for explaining many software functionalities, due to the fact that many concepts exposed in the following paragraphs can be easily arranged at the other devices' simulation interfaces.

Electrostatic actuator – homogeneous Looking at the window provided as a software interface for analysing this device (Figure 9.5), it is possible to focus on the following parameters to be set:

- First, the selection of a material composing the beam, which implies for materials just included in the corresponding *listbox,* the consequent choice of its Young's Modulus (suitable control-buttons are available for deleting existing materials, or their properties, or adding new materials. Information about the default materials and inherent parameters are stored in the two text files: *listbox_materials.txt* and *listbox_materials_paramvalues.tx*).

- Through the others edit-buttons, it is possible to set other mechanical or geometrical parameters as follows: the beam's length, width and thickness, the gap thickness between the beam and the underlying electrode (supposed to be a square-shaped electrode with sides equal to the beam width), the dielectric constant and the viscosity of the material interposed between the beam and the electrode (such a default material is air). It is worth highlighting that a geometrical constraint concerns the beam length-to-width ratio, which is supposed to be greater or equal to 10.

In the right-hand part of the window (see Figure 9.5), the input–output parameters are selectable. In particular, for the input parameters a range of variation is required, allowing us to fix the limits for the input quantities. To produce subsequent plots reporting the output quantities as a function of the input quantities a maximum number of two input parameters can be varied for each simulation. Therefore, three-dimensional graphs can be generated. No constraints are considered for the number of output quantities to be visualized.

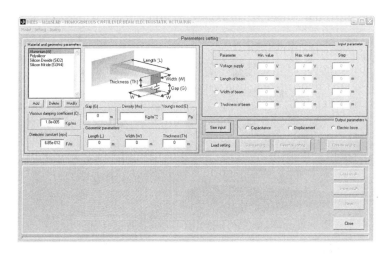

Figure 9.5 Command window for a homogeneous out-of-plane (cantilever) electrostatic actuator

Static or dynamic analysis can be performed, meaning that information on the steady-state response of the system can be obtained (in the case of the sinusoidal regime). Three control-buttons allow us to load (Load setting), save (Save setting) from or in a text file, or reset the simulation parameters (Reset all settings). A final confirmation of the setting parameters' operation is addressed through the control-button 'Confirm setting'. Then, the bottom part of the window is enabled, allowing us to start the simulations and visualize the desired results, as shown in Figure 9.6.

Information regarding the physical features (volume, mass, etc.) and the harmonic response (expected resonance frequency) of the structure are hence summarized in such a section. Through the checkbox (Plot results) the results are visualized, while enabling that the 'Hold on' checkbox iterative simulations can be reported on the same graph.

Results from a simulation can then be saved and recalled when needed. In particular, two files are stored with different extension, a *file.dat*, with the raw data representing the obtained results and a *file.rcd* (Result context data) concerning the simulation setting parameters.

The control-button, 'New', allows for disabling the bottom part of the window and restarts with a new setting of the parameters and simulation. The control-button 'Model' allows us to obtain information on a separated window about the analytical modelling procedures that are implemented in the respective simulation context.

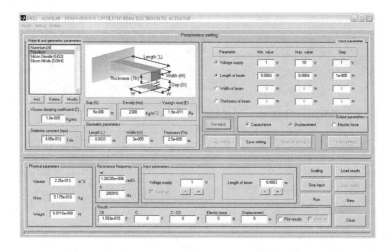

Figure 9.6 Command window for a homogeneous out-of-plane (cantilever) electrostatic actuator. Here, the bottom part of the window has been enabled, allowing for running the simulations and visualizing the desired results

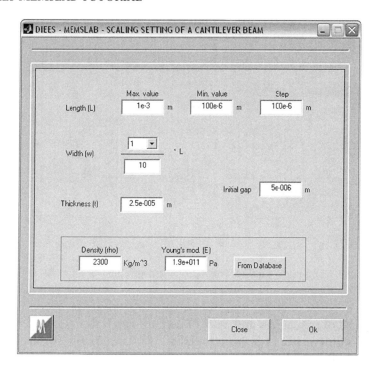

Figure 9.7 Setting command window for analysing scaling in a cantilever device, i.e. an electrostatic actuator

Loading, saving and resetting operations can also be selected from the main menu at the top of the command window (Setting menu).

Such a main menu also includes the 'Model' and 'Scaling' submenus. In the first case, a separated window will report the equations of the considered analytical model. In the second case, another separated command window, such as that shown in Figure 9.7, is displayed. From there, it is possible to set geometrical features and parameter ranges for analysing the scaling issues of the selected device.

The same command window can be operated through the 'Scaling' control-button, at the bottom right-hand side of the main command window.

In the particolar case of an homogeneous cantilever, it is also required to choose the material's physical properties (Young's modulus and density), or as an alternative they can be loaded from the correspondent list-box menu (from the 'Database' control-button).

Thermal actuator – bilayer For cantilever-beam thermal actuators, a well-defined structure has been considered (Figure 9.8). This is

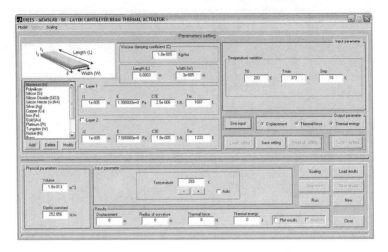

Figure 9.8 Command window for a homogeneous bilayer (cantilever) thermal actuator

composed of two layers and a list of default materials is again provided. The selection of one existing material allows us to fix the corresponding mechanical and physical features, such as the Young's modulus, the thermal expansion coefficient and the melting temperature. For each of the two layers, the thickness have to be set, whereas their lengths (L) and widths (W) are supposed to have equal extensions. The input temperature range and the step of variation then be set, respecting only the limit imposed by the material's melting temperature. The results will report on the tip displacement, the curvature radius, the thermal energy and the thermal force.

Accelerometer – homogeneous If the cantilever beam is supposed to work as an accelerometer, the window reported in Figure 9.9 is visualized. This is very similar to those previously described, and information about the structure geometry, composing material and the viscosity of the fluid in which the device is immersed have to be set. As output results, the tip displacement (in static or dynamic conditions), the mechanical force, elastic constant and the device sensitivity are reported and can be easily plotted. The setting command window for analysing scaling in a cantilever device, i.e. an accelerometer, is presented in Figure 9.10.

Mass sensor – homogeneous In this last case, the cantilever beam has also been considered to be operated as a mass sensor. The variation of

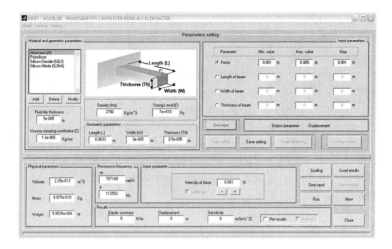

Figure 9.9 Command window for a homogeneous out-of-plane (cantiliver) accelerometer

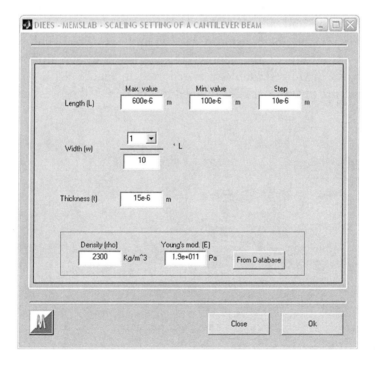

Figure 9.10 Setting command window for analysing scaling in a cantilever device, i.e. an accelerometer

the resonance frequency of the structure, as a function of the variation of the mass (if absorbed by or deposited onto the cantilever surface), is considered as the measuring criteria. The desired range of variation of the mass must be set, whereas the output frequency shift and device sensitivity can be visualized. The control-button 'Auto' allows us to run simulations spanning the range of variation of the mass and producing output results with the frequency shift for each point. The command window for a homogeneous out-of-plane (cantilever) mass sensor and the setting command window for analysing scaling in such a canti-layer device, i.e. a mass sensor, are presented in Figures 9.11 and 9.12, respectively.

Some other concepts related to using the proposed software platform with a further degree of complexity can be clarified, always referring to the cantilever beam as a reference device.

When the homogeneous case is not selected, three different alternative choices allow for considering the same device realized through a standard CMOS process (AMS, 0.8 μm) and wet TMAH micromachining procedures from the front-side, an SOI-based process (described in Chapter 5) and a CUSTOM process in which the number and properties of the materials can be suitably set. The window used for the case-study of the cantilevered electrostatic actuator realized through the CMOS technology is reported in Figure 9.13. A typical thin structure (membrane), is obtained for the depicted process since the

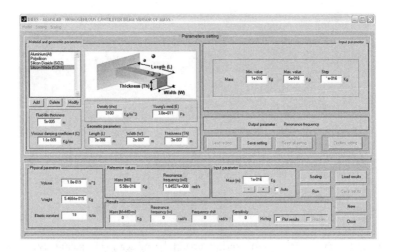

Figure 9.11   Command window for a homogeneous out-of-plane (cantilever) mass sensor

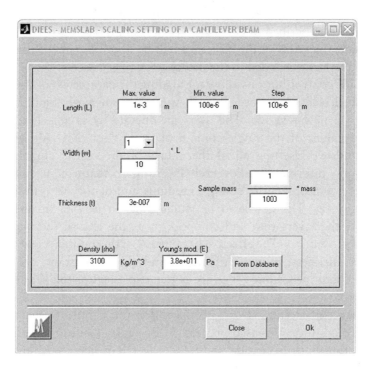

Figure 9.12 Setting command window for analysing scaling in a cantilever device, i.e. a mass sensor

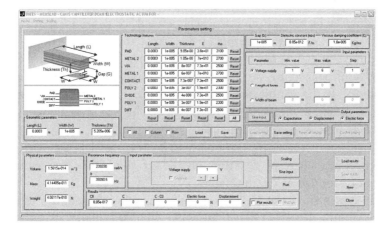

Figure 9.13 Command window for a CMOS out-of-plane (cantilever) electrostatic actuator

micromachining procedures allow for anisotropically etching the silicon substrate, as described in Chapter 2. As a result of such procedures, the equivalent thickness of the structure will be approximately 5 μm, if all of the layers are included in the device.

Suitable control-buttons allow for loading the parameters of the afore-mentioned technology, even if such values can be freely modified and saved in a new text file. The user can operate by saving, loading or deleting values of the single column or 'raw' data. The 'All' control-button allows resetting all of the values of the matrix summarizing the layers' information. For each layer, information concerning the thickness, width, length, Young's modulus and density can, in fact, be separately set. The device to be studied has to be first imagined, in terms of the layers that will compose it, at the end of the microma-chining procedures. The effective parameters of this multilayer structure, such as thickness, moment of inertia and elastic constant, are calcu-lated through procedures that are better explained in the next section (flexures).

Similar considerations can be made for the cantilever beam real-ized through the SOI-based process. In this case, due to the possibility of adopting both front-side and back-side DRIE micromachining, the degrees of freedom during the design phase increase somewhat. Due to the presence of the buried oxide, the final device can, in fact, be composed of the sole c-Si layer (the upper silicon layer) and eventually the thin materials subsequently deposited onto its surface or, they can be very thick if the substrate is suitablly masked from the back-side. The number and types of materials considered are related to the 'custom technology' which has been purposely defined to realize inertial sensors with resistive and optical outputs (reflectance measurements).

The mechanical performances of all of the considered devices are strongly influenced by the micromachining procedures, as shown in Chapter 5. If the cantilever device is again considered, the command window shown in Figure 9.14 will be used. The c-Si layer thickness can be 5 or 15 μm, or as set by the user.

Each layer of the two above-discussed processes has its own name derived from their respective founding design rules (they are just stored and can be loaded as default values).

If, in conclusion, the CUSTOM technology is selected, the user has the maximum degree of freedom in choosing the material properties, both geometrical and physical (also, the layer name can be arbitrarily assigned to create new processes), as shown in Figure 9.15.

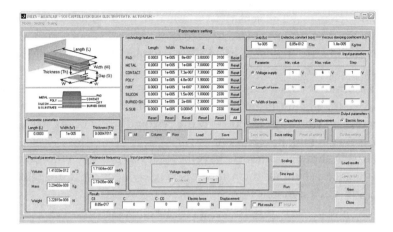

Figure 9.14 Command window for an SOI-based out-of-plane (cantilever) electrostatic actuator

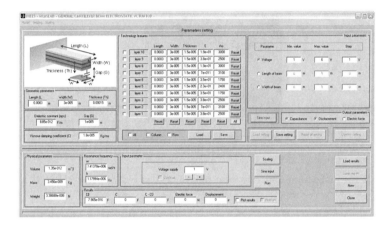

Figure 9.15 Command window for a CUSTOM-based technology out-of-plane (cantilever) electrostatic actuator

9.3.2.2 Flexures

As mentioned in Chapter 5, simple devices composed of a suspended rectangular-shaped proof-mass, anchored by its four corners to the silicon substrate through four symmetrical beams, have been considered to consider other approaches for simulating mechanical sensors. If, as will be discussed below, a symmetric structure is considered, many issues related to adoptable sensing techniques can be solved, due to the fact that the motions of the active surfaces can be considered to be normal to the device plane. The sensitivity attainable with optical or capacitive sensing can then be maximized, whereas many issues concerning analytical modelling are also simplified. In particular, in the case of optical sensing performed through interferometric strategies, the high degree of parallelism between the fronting surfaces (usually realizing optical cavities) can be better insured due to the symmetry of the system which guarantees a symmetric reaction to the residual stress occurring during the micromachining procedures.

'Lumped' parameter models have been developed for bulk micromachined devices, dividing them into discrete elements that are modelled using rigid-body dynamics.

In particular, two different shapes have been considered for the sustaining springs:

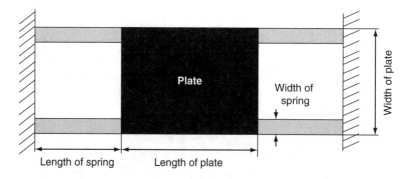

Figure 9.16 Schematic top-view of a reference 'flexure' device with fixed springs

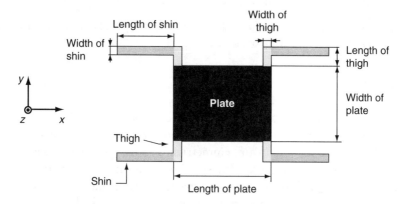

Figure 9.17 Schematic top-view of a reference device with 'crab-leg' flexures

- Fixed springs (Figure 9.16);
- 'Crab-leg' springs (Figure 9.17).

At the same time, three different opportunities are offered by the simulator for the technology choice, meaning choosing materials that compose the mechanical elements (beams, plate, fingers, etc.):

- Homogeneous;
- CMOS (AMS, $0.8\,\mu m$);
- SOI-based processes with front- and back-side DRIE etching.

For each of the previous structures, static and dynamic analysis can be made by running suitable procedures that allow for evaluating the elastic constants (an important issue for multilayered devices), the damping factor, the proof-mass and so on.

Finally, the displacement of the structure following an external mechanical stimulus can be evaluated.

The analysis of such micromechanical freestanding structures has been done by using an energy method to derive analytical formulae for linear spring constants.

Due to the hypothetical symmetry of the systems, important for determining boundary conditions, it is possible to consider only one spring and then to multiply by four the obtained spring constants.

Linear equations for the spring constants are derived using energy methods. A force F_i (or moment M_i) is applied to the free end of the spring, in the direction of interest and then the displacement is calculated from $\delta_i = k_i/F_i$ (with i being x, y or z).

In these calculations, different boundary conditions, with hypothetical symmetrical systems, are applied for the different modes of deformation of the spring. Then, from the second Castigliano theorem , the partial derivative of the strain energy of a linear structure, U, with respect to a given load, is equal to the displacement at the point of application of the load, δ. Only displacements resulting from bending and torsion are considered in the analysis, while secondary effects as deformations from shear, beam elongation and shortening are neglected.

The expressions that we have considered for the total strain energy in the case of fixed springs and 'crab-leg' springs are, respectively:

$$U = \sum_{i=1}^{N} \int_0^{L_i} \frac{M_i(\xi)^2}{2EI_i}; \quad U = \sum_{i=1}^{N} \left(\int_0^{L_i} \frac{M_i(\xi)^2}{2EI_i} d\xi + \int_0^{L_i} \frac{T_i(\xi)^2}{2GJ} d\xi \right) \quad (9.11)$$

where M_i is the bending moment transmitted through the ith beam, L_i is the length of the ith beam in the spring, E is the Young's modulus of the structural material (if homogeneous) and I_i is the moment of inertia of the ith beam, about the reference axis, whereas T is the torsion, G is the torsional modulus of elasticity and J is the torsional constant, considered in the case of non-linear spring-constant evaluation (as in the case of 'crab-leg' springs).

Details of the analytical approach for calculating the elastic constants of the considered two devices can be found in Fedder (1994) and Iyer *et al.* (1999).

The final expressions for the elastic constants of fixed and 'crab-leg' springs along the z-axis, the axis normal to the device plane, are reported in equations (9.12) and (9.13).

The simulator also gives an opportunity for calculating the elastic constants along the other two directions, derived with the same analytical method. The elastic constant along the z-direction for the fixed beam is:

$$k_{z,\text{beam}} = \frac{12 E_n I_{ny}}{L^3} \tag{9.12}$$

The spring constant of the flexure is four times the beam's elastic constant in the z-direction:

$$k_z = \frac{48 E_n I_{ny}}{L^3} \tag{9.13}$$

where E_n is the normalized Young's modulus in a layered beam and I_{ny} is the equivalent moment of inertia. The elastic constants along the direction x and y can be easily determined.

Regarding the elastic constant of the 'crab-leg' flexure, this can be expressed as follows:

$$K_z = \frac{4F_z}{\delta_z} = \frac{48 H_{Ea} H_{Eb} \left(H_{gb} L_t + H_{Ea} L_s \right) \left(H_{Eb} L_t + H_{ga} L_s \right)}{\left(\begin{array}{l} H_{Eb}^2 H_{gb} L_t^5 + 4 H_{Ea} H_{Eb}^2 L_t^2 L_s + H_{Eb} H_{ga} H_{gb} L_t^4 L_s + 4 H_{Ea} H_{Eb} H_{ga} L_t^3 L_s^2 + \\ 4 H_{Ea} H_{Eb} H_{gb} L_t^2 L_s^3 + 4 H_{Ea}^2 H_{Eb} L_t L_s^2 + H_{Ea} H_{ga} H_{gb} L_t L_s^4 + 4 H_{Ea}^2 H_{ga} L_s^5 \end{array} \right)} \tag{9.14}$$

where $H_{Ea} \equiv EI_{x,a}$ and $H_{Eb} \equiv EI_{x,b}$, with $I_{x,a}$ and $I_{x,b}$ the moments of inertia of the thigh and shin with respect to the x-axis; $H_{ga} \equiv GJ_a$ and $H_{gb} \equiv GJ_b$. The torsional modulus is related to the Young's modulus and the Poisson ratio, ν, as follows:

$$G = \frac{E}{2(1+\nu)} \tag{9.15}$$

The torsional constant for a beam of rectangular cross-section is given by:

$$J = \frac{1}{3} t^3 w \left[1 - \frac{192}{\pi^5} \frac{t}{w} \sum_{i=1, i, \text{odd}}^{\infty} \frac{1}{i^5} \tan h \left(\frac{i\pi w}{2t} \right) \right] \tag{9.16}$$

where $t < w$. An alternative approximate expression has been used, with $t \ll w$:

$$J = \frac{1}{3} t^3 w \left(1 - 0.630 \frac{t}{w} \right) \tag{9.17}$$

Finally, the following procedure, previously mentioned in Chapter 5, has been implemented to calculate the moments of inertia of layered structures:

❖ Calculate the normalized widths:

$$w_{i\_\text{norm}} = w_i \left(\frac{E_i}{E_{\text{max}}} \right) \tag{9.18}$$

❖ Calculate the normalized sections:

$$s_{i\_\text{norm}} = w_{i\_\text{norm}} t_i \tag{9.19}$$

❖ Calculate the vertical location of the neutral axis:

$$h_i = \sum_{k=1}^{i-1} t_k + \frac{t_i}{2} \tag{9.20a}$$

$$h_n = \frac{\sum_{i=1}^{N} \left(s_{i\_\text{norm}} h_i \right)}{\sum_{i=1}^{N} s_{i\_\text{norm}}} \tag{9.20b}$$

❖ Calculate the equivalent moments of inertia:

$$I_{i\_\text{norm}} = I_i + s_{i\_\text{norm}} \left(h_n - h_i \right)^2 \tag{9.21}$$

$$I_n = \sum_{i=1}^{N} I_{i\_\text{norm}} \tag{9.22}$$

Algorithms for calculating the parameters expressed in equations (9.13–9.21) have been implemented into the 'Matlab' code. When a 'flexure' device is selected, it can be analysed both as an accelerometer or a mass sensor. The following selection will regard the shape of the sustaining beams. If the fixed–fixed configuration is selected and a homogeneous configuration is also chosen, the command window reported in Figure 9.18 is displayed and can be operated. Geometrical (thickness, width and length) and physical (Young's modulus and density of materials) properties of the beams can be typed, or loaded from a file (\\flexures\CMOS).

Several control-buttons allow for setting the device features, whereas the operating mode is very similar to that described above for simulating the cantilever beam. Additional information has to be provided only for the geometrical features of the central part of the device, named as the 'plate'.

If instead the 'crab-leg' flexure is selected, coupled with the homogeneous material composition, the command window shown in Figure 9.19 is displayed. Among the geometrical features to be specified, two new parameters must be set, being the arms of the 'crab-leg' suspensions. These are, respectively, the 'thigh' (segment anchored to the plate) and 'shin' (segment anchored to the substrate). The torsional constant is automatically derived from equation (9.19) or can be set by the user.

On the other hand, if a different technology is selected, the command windows shown in Figures 9.20 and 9.21, for a fixed–fixed CMOS

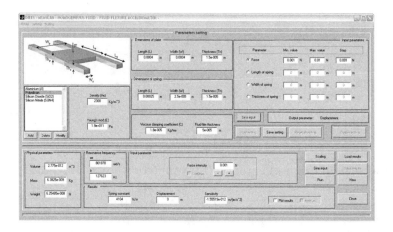

Figure 9.18 Command window for a homogeneous, fixed–fixed flexure used as an accelerometer

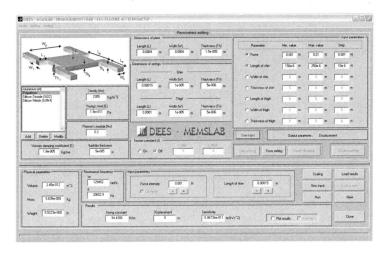

Figure 9.19 Command window for a homogeneous, 'crab-leg' flexure used as an accelerometer

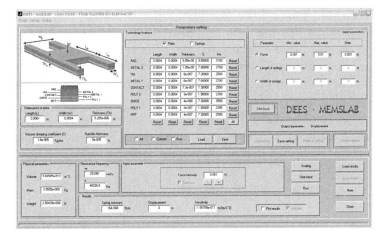

Figure 9.20 Command window for a CMOS, fixed–fixed flexure used as an accelerometer

flexure (accelerometer) and a 'crab-leg' SOI flexure (accelerometer), respectively, will be displayed.

It is worth while to highlight that for the CMOS technology and the concerned micromachining procedures, the substrate is never included in the parameters list since the released structures are always very thin (a few microns).

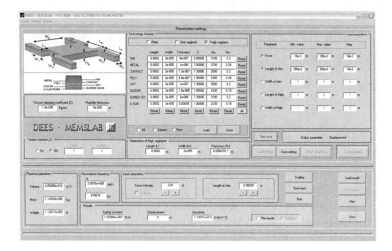

Figure 9.21 Command window for an SOI-based, crab-leg flexure used as an accelerometer

On the other hand, in the case of SOI-based devices, it is realistic to conceive devices in which the c-si layer (being preferably 5 or 15 μm thick) is included in the beams and the plate areas, while the substrate can be removed (or not) from the bottom part of the plate area, to realize thick and heavy proof-masses (as required to improve the sensitivity for z-axis accelerometers). Differences in the cross-sections of such basic devices can be observed in Figure 9.22.

Moreover, for the proposed micromechanical plates the simulator is structured as described in the following section.

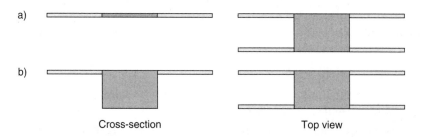

Cross-section　　　　　　　　Top view

Figure 9.22 Schematics of (a) a cross-sectional and (b) a top view of two reference SOI devices. In the former case, with a suitable back-side bulk micromachining procedure a larger proof-mass can be obtained

9.3.2.3   Interdigitated Devices

As a last possibility, from the main command window it is possible to select options for running simulations on interdigitated actuators or sensors, through the radio-button 'Interdigitated'.

In particular, for both the categories of devices the same options are available for selecting the realization process and the geometrical features. The geometrical and physical parameters of interdigital fingers are the additional information to be provided in this case. Transverse or lateral actuators can be selected, whereas accelerometers with capacitive outputs are considered as sensors, and a lateral or transverse read-out is implemented.

It is also possible to select the suspensions' shapes, among 'crab-leg' or fixed–fixed systems, as in the case of flexures. This will influence the elastic constants of the structures and then the final results for both actuators or sensors.

In the case of transverse combs, an ulterior distinction can be made between *Single finger* and *Double finger*, referring to the number of stator fingers interposed between the two rotor fingers. For the single-finger configuration, it is necessary to choose between a differential or a single-ended read-out of the capacitance variation. As an example, the command window displayed for running simulations on transverse comb differential actuators, with fixed–fixed suspensions and single-finger configurations, is reported in Figure 9.23.

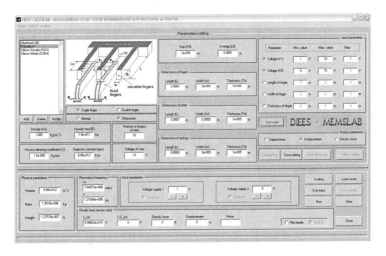

Figure 9.23   Command window for a homogeneous interdigitated, fixed–fixed actuator

The image displayed in Figure 9.23 shows only a portion of the device which is made by a central part (plate) where the rotor fingers are anchored on two sides. The features of such a central part can be set by using the 'Plate settings' check-box button on the same window.

The settings of the parameters require the definition of material properties, including features of the material surrounding the fingers (usually the viscosity of air). For simplicity, the rotor and stator fingers are assumed to have equal geometrical features. In addition, the distance (gap) and the overlap between the fingers must be set. For the sake of clarity, the meanings of such latter parameters are reported in Chapter 2. From an electrical point of view, the potentials at each electrode array have to be fixed. It is worthwhile to observe that for both lateral and transverse combs the 'gap' between a finger and the opposite part (named as the frontal gap, F_{gap}) is not 'typed' at the simulator, but determined from other data (length and overlap). Once the geometrical and physical properties of the rotor fingers are set, the number of stator fingers is automatically determined by selecting lateral or transverse combs.

If the other available technologies are selected, the displayed command window is usually similar to that reported in Figure 9.24, in which a CMOS, fixed–fixed, lateral comb actuator is analysed.

Then, static or dynamic analysis can be carried out. Displacements, electrostatic forces and capacitances values can be iteratively plotted, acting on the 'slider buttons' that, in this case, is in the 'voltage supply'

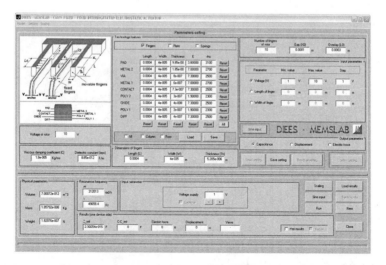

Figure 9.24 Command window for a CMOS interdigitated, fixed–fixed actuator

section. In each of these cases, the results window will allow us to plot and save results for the total and partial capacitances between the fingers, the total and partial electrostatic forces and the displacements obtained for the actuators.

9.4 CONCLUSIONS

Coherently with the purposes of basic courses on MEMS, the proposed simulator will be extended year-after-year. In fact, even though advanced academic courses may also include powerful CADs for MEMS' modelling and design, the proposed instrument obtains familiarity with the first concepts on such topics, also giving the opportunity to make variations to the code (since 'Matlab' can be retained an open-source for academic use).

To this purpose, the guidelines that the authors suggest to follow for extending the software tools are:

- Include cantilever beams as a third reference structure, operated as a passive or resonant mechanical sensor and actuator, with capacitive or thermal inputs or outputs.
- Extend the libraries for technologies (meaning that, except for CMOS or SOI, some other technologies could be included). Alternately, a general-purpose user interface, without any link to a specific technology, could be created.
- Extend the libraries for mechanical structures. Analytical models of the following structures could be included: rotating structures operating, for example, as RF switches; vibrating or rotating structures operating as capacitive-plate gyroscopes; two- or three-axis interdigitated combs operating as interdigitated capacitive accelerometers and gyroscopes, and so on.
- Extend the libraries for sustaining shapes of the springs. Many other solutions allow us to optimize several different aspects inherent to mechanical 'cross-talk', compliance in a specific directions, etc.
- Include models of mass sensors or pressure sensors, both for cantilever beams or plates, governing also the behaviour of many MEMS used for chemical or biological applications (independently from resonant or static operative modes).

Although, it is probably not worthwhile to excessively extend the described simulator, there is more than one reason to consider such software instruments for gaining confidence with microelectromechanical systems, since often a deep understanding and correct usuage of more and more powerful CADs are not guaranteed over a short time, so preventing us to appreciate and suitably take into account all of the aspects of such a highly interdisciplinary 'microworld'.

REFERENCES

G.K. Fedder (1994). *Simulation of Microelectromechanical Systems*, Ph.D. Thesis, EECS Department, University of California at Berkeley, CA, USA.

S. Iyer, Y. Zhou and T. Mukherjee (1999). Analytical Modeling of Cross-axis Coupling in Micromechanical Springs, in *Proceedings of MEMS'99*, San Juan, Puerto Rico, April 19–21, pp. 632–635.

10

Concluding Remarks

Scaling in micro- and nano-electromechanical systems has been addressed and adopted in this work as a methodology and philosophy for the analysis, modelling, simulation, design and characterization of MEMS.

The concept of scaling for electromechanical systems has been introduced and analysed while the effects induced in the device structures, in the physical laws and in the technologies and performances have been taken into account.

Among all of the possibilities, the attention here has been focused on semiconductor-based electromechanical systems because of their interesting features, such as low-cost, large-volume fabrication and integration with electronics.

Moving from the macro- to the microscales requires a review of the models and the approximations commonly used on the larger scale. In fact, while the basic physical phenomena governing the behaviour of a given device are the same, their relative influences change – then an accurate analysis of scaling is required.

By further scaling object sizes towards the nanoscale, some interesting effects may arise. In particular, although objects are still big enough to be described with classical physics laws, some macroscopic quantum effects become observable. Thus, nanoscale devices are an interesting possibility to explore and study a novel class of phenomena.

A very important advantage of miniaturization is that it provides the access to a wide spectrum of novel, unforeseen applicative fields.

Scaling Issues and Design of MEMS   S. Baglio, S. Castorina and N. Savalli
© 2007 John Wiley & Sons, Ltd

Many aspects of the modelling, design, fabrication and characterization of MEMS have been reviewed by taking into account scaling principles, and they have been presented here together with some concrete realizations. Several different fields have been addressed, with different levels of details and development.

In particular, some introductory concepts about the scaling of microsystems, together with the basic ideas of this work, have been introduced.

The basic scaling concepts have been applied and detailed to several concrete applications. These include microactuators, thermal, magnetic and mechanical sensors, and energy sources.

Then, the issues related to scaling, and the technologies used for their fabrication, have been summarized to describe the potential implementation of autonomous microrobots. Moreover, microrobots have been proposed as a paradigm for the study of scaling laws and their effects in the design of microsystems.

Finally, an overview of nano-electromechanical systems has been provided as a scenario of the natural evolution for the further scaling of some of the devices previously discussed. Although not exhaustively, some of the most important examples of current NEMS have been reported. Semiconductor-based NEMS are still in their early stages; however, they show enough potential to gain a lot of interest as estalished MEMS and perhaps even more so.

Molecular-scale electronic and electromechanical devices are promising approaches, but they are still premature. However, since much research effort has been devoted to these branches of nanotechnology, important innovations may soon arise.

A kind of 'roadmap', going from miniaturized systems to micro- and nanometric devices has been followed in this work. Both methodological issues and experimental and applicative examples have been 'touched' by this trail, aiming to give a deeper understanding of the subject.

Index